Phase Equilibria Diagrams of High Temperature Non-oxide Ceramics

Zhenkun Huang · Laner Wu

Phase Equilibria Diagrams of High Temperature Non-oxide Ceramics

Springer

Zhenkun Huang
Materials Science and Engineering
Beifang University of Nationalities
Yinchuan, Ningxia
China

Laner Wu
Materials Science and Engineering
Beifang University of Nationalities
Yinchuan, Ningxia
China

ISBN 978-981-13-4422-0   ISBN 978-981-13-0463-7   (eBook)
https://doi.org/10.1007/978-981-13-0463-7

This Springer imprint is published by the registered company Springer Nature Singapore Pte Ltd. part of Springer Nature
The registered company address is: 152 Beach Road, #21-01/04 Gateway East, Singapore 189721, Singapore

Foreword

The progress of civilization has largely encompassed the usage, consumption, and creation of materials. Nowadays, the advancement of new technologies has often based on the discovery of new materials and the ability of improving the physical, chemical, and biological properties of materials. Existing materials can no longer satisfy the requirement of the rapid evolution and revolution of modern technologies; simple synthetic materials, in many cases, are not able to meet the rigid demands of multifunctional and high-performance devices and systems, particularly for the applications under extreme conditions. The materials with complex composition or integrated multiple distinctive structures away from thermodynamic equilibrium state have become a new research direction, due to the promising properties of these materials. The study of such materials requires the researchers, including chemists, materials scientists, and engineers, to have a better understanding of the fundamentals of materials science: phase diagram, also known as phase equilibria diagram, the chart to show where the distinct phases exist and coexist at thermodynamic equilibrium, and thus plays a crucial role in the field of materials research.

Phase diagrams serve as a big database for materials scientists and engineers and a guide or a map in the search for new materials and materials away from thermodynamic equilibrium. For example, early day's purification of silicon semiconductor was greatly benefitted from the phase diagram, and the synthetic diamond was grown under the thermodynamic equilibrium. The growth of silicon nanorods or nanowires, carbon nanotubes, and graphenes by vapor–liquid–solid (VLS) and solid–liquid–solid (SLS) methods is yet another example that phase diagrams are a basic and powerful tool in guiding the search for materials with new structures and properties. Since the first volume of phase diagrams for ceramists was published in 1956, many volumes followed, now consisting of a collection of 20,000 diagrams.

High temperature non-oxides have found a lot of applications in the fields of aerospace, aviation, oil extraction, machining, biomedical, and weaponry. Due to their inherent thermal and chemical stability at extreme conditions, synthesis, processing, and fabrication of such non-oxide ceramics and their components are challenging and often require extremely high temperatures for a prolonged period of time in inert gas, sometimes with the assistance of high pressure as well. At a high temperature, some constituent elements may escape through evaporation or diffusion; impurities in gas and crucibles may diffuse in and be incorporated to the crystal lattice or form parasitic phases. In addition, various additives are commonly intentionally introduced to the system to circumvent the stringent processing conditions and to tailor the desired chemical and physical properties of non-oxide ceramics, making the materials absolutely complex and creating extra work for the subsequent characterization. So the study on phase diagrams of high temperature non-oxides are extremely valuable, yet technically challenging and time-consuming.

Phase Equilibria Diagrams of High Temperature Non-oxide Ceramics systematically compiled and summarized the recent accomplishment in the study of high temperature non-oxide ceramic phase diagrams with a comprehensive collection of the diagrams. This authoritative book will surely equip scientists and engineers with a holistic understanding of the existence and coexistence of various phases and enable them to design viable processing approaches, to select applicable materials for various applications, and to predict the potential degradation under extreme conditions.

The editors of this book are the acclaimed experts in the field of high temperature non-oxide ceramics, particularly in the study of phase diagrams of those materials. Professor Zhenkun Huang, together with his former colleagues, Prof. Weiying Sun in particular, has made an extraordinary contribution to the study of phase diagrams of high temperature non-oxide ceramics in the past 4 decades at Shanghai Institute of Ceramics of Chinese Academy of Sciences, Max Planck Institute, University of Newcastle, University of Michigan, and most recently at North University of Nationalities with Prof. LanEr Wu, the co-editor of this book.

Seattle, WA, USA   Guozhong Cao
May 2017   Boeing-Steiner Professor
                                                                      University of Washington

Preface

Phase diagram of ceramics is a diagrammatic description of phase equilibrium relations among multiple components in ceramic systems through physicochemical reaction under specific temperature and pressure. It accumulates the results of experimental research of materials, and it is one of the foundations of materials' development. Nowadays, high temperature non-oxide ceramics, such as borides, carbides, and nitrides, have been widely studied, and they have been applied in different technical fields, not only for civil use but also in important industries like military, aviation, aerospace, etc.

This book mainly includes the non-oxides ceramics phase diagrams based on borides, carbides, nitrides of silicon, aluminum, and transition metals in groups IV, V, and VI. The contents of the book include four chapters as follows:

Chapter 1: Si_3N_4 Ceramics Systems.
Chapter 2: SiC-Dominated Ceramics Systems.
Chapter 3: AlN-based Ceramics Systems.
Chapter 4: Ultrahigh-Temperature Ceramics (UHTCs) Systems.

Differing from common oxides and halides, most of the non-oxides possess the strong covalence and have supreme properties of high hardness, high strength, and high temperature resistance. Many of these compounds have no congruent melting point and decompose at high temperature.

Consequently, subsolidus phase diagrams are constructed via experiments of solid-state reactions and sintering. Samples are usually analyzed by means of X-Ray Diffraction (XRD), Scanning Electron Microscope (SEM), electronic probe, metalloscope, Thermal Analyzer (TG), Differential Thermal Analyzer (DTA), etc.

Subsolidus phase diagrams gained under the solidus temperature are usually produced through experiments of the solid-state reaction. Since some of these compound systems are difficult to identify their phase relations in equilibrium through experiments, the calculated phase diagrams could be obtained by means of thermodynamic calculation. Most of the phase diagrams collected in this book are experimental phase diagrams with some relevant calculated phase diagrams being gathered together.

As a fundamental theory for a given multiphase/multicomponent system, Josiah Willard Gibbs proposed the phase rule in 1876 to relate the total degrees of freedom (f) of the system, the number of phases (Φ) and the number of components (c), simply as $f = c - \Phi + 2$, where the number 2 is devoted to number of variable external factors of temperature and pressure. Condensed state system is the system only contains liquids and/or solids, in which the influences of gas state are ignored. The external factor taken into account is only temperature. Therefore, the phase rule is replaced by $f = c - \Phi + 1$. The most of equilibrium phase diagrams comply with the phase rule, except some "behaviors" phase diagrams which are difficult to achieve phase equilibrium.

The phase rule will help readers to understand the equilibrium phase diagrams. Using these compounds as the end-member components, the phase equilibria diagrams of the binary, ternary, quaternary, pentabasic, and even hexahydric (Jänecke prism) systems have been presented.

There are 301 figures involving about 150 material systems in the book. Most of the selected phase diagrams are taken from ACerS-NIST Phase Equilibria Diagrams PC Database Version 4.0 (or Version 3.4) (http://ceramics.org/publications-and-resources/phase-equilibria-diagrams), or be called PED in this book. Some of the newly published works by the authors' team are also presented in this book. The research achievements introduced in the book include the author Prof. Zhenkun Huang's works accomplished in different "phase equilibrium diagrams" groups, including Powder-Metallurgic Laboratory (PML) Max Planck Institute, Germany, Shanghai Institute of Ceramics, Chinese Academy of Sciences with colleague Prof. Weiying Sun, Department of Material Science and Engineering, University of Michigan, as well as North University of Nationalities, P. R. China, in recent 10 years. The appendix attached at the end of the book includes six structure data tables. Table A.1 "Melting point and structure of some compounds in Y–Si–Al–O–N system", Table A.2 "Eutectic temperature and composition in some nitrogen-ceramic systems", Table A.3 "Compounds of nitrogen-containing metal aluminosilicates", and Table A.4 "Compounds of some nitrogen-containing rare earth aluminosilicates" are collected from nitrogen ceramics series reports in the 1970–1990s, which will be helpful for people who are doing research in nitrogen ceramics and other high temperature non-oxide ceramics. Table A.5 "Crystal chemistry data of some compounds with tetrahedral [MX_4] structure" and Table A.6 "Crystal structure, lattice parameter, density, and melting point of Ultrahigh-Temperature Materials (UHTMs)" will be helpful to understand the high temperature reaction, phase compatibility, and phase relations of the different compounds.

High temperature non-oxide ceramics are playing an important role in structure ceramics field, especially since 70s of the twentieth century. From nitrogen ceramics (basically Si_3N_4) to SiC, AlN, and even ultrahigh-temperature ceramics as diborides, monocarbides, and mononitrides, a series of modern ceramics of high temperature non-oxides have been developed. The contents of this book have been sorted in the order of Si_3N_4 ceramics systems, SiC-dominated ceramics systems, AlN-based ceramics systems, and UHTCs systems successively. Parts of V and VI group transition metals non-oxide high temperature ceramics systems (not UHTC) are also included. Regarding to each figure selected from PED database, a relevant PED number could be found on the up left corner of the figure. Readers can easily follow the number to search the original documents in the database. To avoid the confusion of the figure tags, use upper case, A, B, C,..., for the figure tags in PED; use lower case, a, b, c..., for the figure tags selected from PED in the book.

Editor of this book, Prof. Zhenkun Huang, is a former research scientist of Shanghai Ceramics Institute, Chinese Academy of Science, and was an Exchange Scholar of Max Planck Institute, Germany (1980–1982), and was a Visiting Research Scientist of Michigan University. Co-editor Prof. Laner Wu from North University of Nationalities, P. R. China has been working together with Prof. Huang in recent 15 years. Contributions for this book also include teachers and postgraduate students from North University of Nationalities, P. R. China: Zhenxia Yuan, Wengao Pan, Wanxiu Hai, Yong Jiang, and Yuhong Chen; Limeng Liu made great contribution to English revision and helpful comments. Dr. Jie Yang made some English revision too.

We express our sincere thanks to the authors of the original works, to the editors of the PED commentary, and to American Ceramics Society for permission of reprinting the figures from PED PC database. We also thank the support from National Natural Science Fund of China.

Ann Arbor, MI, USA   Zhenkun Huang
June 2017

The original version of the book was revised:
For detailed information please see Erratum.
The erratum to the book is available at https://
doi.org/10.1007/978-981-13-0463-7_5

Contents

1 **Si₃N₄ Ceramics Systems** 1
   1.1 Si$_3$N$_4$–M$_x$X$_y$ Systems 1
      1.1.1 Si$_3$N$_4$–SiO$_2$ 2
      1.1.2 Si$_3$N$_4$–Y$_2$O$_3$ 3
      1.1.3 Si$_3$N$_4$–Y$_3$Al$_5$O$_{12}$ 4
      1.1.4 Si$_3$N$_4$–Y$_2$Si$_2$O$_7$ 4
      1.1.5 Si$_3$N$_4$–Y$_3$Al$_5$O$_{12}$–Y$_2$Si$_2$O$_7$ 5
      1.1.6 Si$_3$N$_4$–SiO$_2$–La$_2$O$_3$ 6
      1.1.7 Si$_3$N$_4$–SiO$_2$–Ce$_2$O$_3$ 7
      1.1.8 Si$_3$N$_4$–SiO$_2$–Gd$_2$O$_3$ 8
      1.1.9 Si$_3$N$_4$–SiO$_2$–Y$_2$O$_3$ 8
      1.1.10 Si$_3$N$_4$–La$_2$O$_3$–Y$_2$O$_3$ 10
      1.1.11 Si$_3$N$_4$–SiO$_2$–Th$_3$N$_4$–ThO$_2$ 11
      1.1.12 Si$_3$N$_4$–SiO$_2$–AlN–Al$_2$O$_3$–ZrN–ZrO$_2$ 11
      1.1.13 Si$_3$N$_4$–SiO$_2$–ZrN–ZrO$_2$–Y$_2$O$_3$ 14
      1.1.14 Si$_3$N$_4$–SiO$_2$–Mg$_3$N$_2$–MgO 14
      1.1.15 Si$_3$N$_4$–Si$_2$ON$_2$–Mg$_2$SiO$_4$ 16
      1.1.16 Si$_3$N$_4$–Ca$_3$N$_2$–Mg$_3$N$_2$–MgO–SiO$_2$ (+CaO) 18
      1.1.17 Si$_3$N$_4$–SiO$_2$–AlN–Al$_2$O$_3$–Ca$_3$N$_2$–CaO 19
      1.1.18 [Si$_3$N$_4$(60 Mol%)–SiO$_2$]–[Si$_3$N$_4$(60 Mol%)–Al$_2$O$_3$]–
             [Si$_3$N$_4$(60 Mol%)–Y$_2$O$_3$] 21
   1.2 β–SiAlON Systems 21
      1.2.1 Si$_3$N$_4$–SiO$_2$–AlN–Al$_2$O$_3$ (1) 22
      1.2.2 Si$_3$N$_4$–SiO$_2$–AlN–Al$_2$O$_3$ (2) 22
      1.2.3 Si$_3$N$_4$–Li$_2$O–Al$_2$O$_3$ 24
      1.2.4 Si$_3$N$_4$–SiO$_2$–AlN–Al$_2$O$_3$–Be$_3$N$_2$–BeO 26
      1.2.5 Si$_3$N$_4$–Al$_2$O$_3$–MgO 27
      1.2.6 Si$_3$N$_4$–SiO$_2$–AlN–Al$_2$O$_3$–MgO 29
      1.2.7 Ln$_2$O$_3$–Si$_3$N$_4$–AlN–Al$_2$O$_3$ (Ln = Nd, Sm) 29

      1.2.8 Si_3N_4–AlN–Al_2O_3–Y_2O_3 29

      1.2.9 Si_3N_4–SiO_2–AlN–Al_2O_3–Y_2O_3 32

      1.2.10 Si_3N_4–$Y_{12}Si_5O_{28}$–$Y_3Al_5O_{12}$–Al_3O_3N 34

      1.2.11 Si_3N_4–4(YN)–4(AlN)–2(Al_2O_3)–3(SiO_2)–2(Y_2O_3) 36

      1.2.12 Si_3N_4–Al_4N_4–Y_4O_6 37

  1.3   α–SiAlON System 38

      1.3.1 Si_3N_4–AlN–R_2O_3 System 38

      1.3.2 Si_3N_4–AlN–M_2O, –M'O System 38

  1.4 O'–SiAlONs 42

      1.4.1 Si_2ON_2–Al_2O_3–M_xO_y 42

  1.5   M'(R)–SiAlONs System 44

      1.5.1 Si_3N_4–R_2O_3–Al_2O_3 (M'(R)–SiAlONs) 44

  1.6   AlSiONs (SiAlON Polytypoids). 46

      1.6.1   SiAlONs Polytypoids 46

      1.6.2 Si_3N_4–$(Be_3N_2)_2$–$(BeO)_6$–$(SiO_2)_3$ 47

  References 48

2 SiC-Dominated Ceramics Systems. 51

  2.1   SiC–SiO_2 51

  2.2   SiC–Si_2ON_2 52

  2.3   SiC–Si_3N_4 53

  2.4   SiC–Si_3N_4–SiO_2 54

  2.5   SiC–Al_4C_3–Be_2C 54

  2.6   SiC–CrB_2 56

  2.7 Al_4C_3–4(SiC)–B_4C 57

  2.8   SiC–SiO_2–Y_2O_3 58

  2.9   SiC–Si_3N_4–R_2O_3 (R = La, Gd, Y) 59

  2.10 SiC–Al_2O_3–SiO_2–Pr_2O_3 59

  2.11 SiC–Al_2O_3–SiO_2–Nd_2O_3 62

  2.12 SiC–Al_2O_3–SiO_2–Gd_2O_3 63

  2.13 SiC–Al_2O_3–SiO_2–Yb_2O_3 64

  2.14 SiC–Al_2O_3–SiO_2–Y_2O_3 65

  2.15 SiC–SiO_2–Al_2O_3–ZrO_2 and SiC–SiO_2–Al_2O_3–MgO 65

  References 67

3 AlN-Based Ceramics Systems 69

  3.1   Al–Si–N–C–O (AlN–Al_4C_3–Al_2O_3–Si_3N_4–SiC–SiO_2) 69

  3.2   AlN–Al_2O_3. 70

  3.3 Al_2O_3–Al_4C_3 72

  3.4   AlN–Al_4C_3. 75

  3.5 Al_2O_3–Al_4C_3–AlN 75

  3.6   AlN–SiC 75

  3.7   AlN–Al_2OC–SiC 76

  3.8 Al_4C_3–SiC 76

3.9   AlN–Al$_4$C$_3$–Si$_3$N$_4$–SiC System 80
3.10 Experimental Phase Diagram of AlN–Al$_4$C$_3$–SiC System 81
3.11 Calculated Phase Diagram of AlN–Al$_4$C$_3$–SiC System 82
3.12 AlN–GaN, GaN–InN, and AlN–InN 85
3.13 AlN–Eu$_2$O$_3$ and AlN–Nd$_2$O$_3$ 85
3.14 AlN–Y$_2$O$_3$ 86
3.15 2(Mg$_3$N$_2$)–4(AlN)–2(Al$_2$O$_3$)–6(MgO) 87
3.16 AlN–Al$_2$O$_3$–Mg$_3$N$_2$–MgO 87
3.17 2(Mg$_3$N$_2$)–4(AlN)–2(Al$_2$O$_3$)–6(MgO) 88
3.18 2(Ca$_3$N$_2$)–4(AlN)–2(Al$_2$O$_3$)–6(CaO) 89
3.19 AlN–Al$_2$O$_3$–R$_2$O$_3$ (R = Ce,Pr,Nd,Sm) 90
3.20 Nd–Al–Si–O–N Jänecke Prism Phase Diagram 93
3.21 AlN–Be$_3$N$_2$–Si$_3$N$_4$ 94
3.22 AlN–SiC–R$_2$O$_3$ (R: Pr, Nd, Sm,Gd,Yb,Y) 97
References 100

4 **Ultrahigh-Temperature Ceramics (UHTCs) Systems** 103
4.1   Introduction 103
4.2   TiB$_2$–TiC$_{0.9}$ and TiB$_2$–B$_4$C 104
4.3   TiC$_{0.95}$–TiB$_2$ 104
4.4   TiC–ZrC 105
4.5   TiC–HfC 107
4.6   TiC$_{1-x}$–HfC$_{1-x}$ 108
4.7   VC$_{0.88}$–TiC 108
4.8   TiC–NbC 112
4.9   TiC–TaC 112
4.10 TiB$_2$–GdB$_6$ 112
4.11 TiB$_2$–TiN$_x$ 113
4.12 ZrC$_{0.88}$–ZrB$_2$ and ZrB$_2$–B$_4$C 115
4.13 ZrC–ZrB$_2$ 117
4.14 ZrC–VC$_{0.88}$ 118
4.15 ZrC–NbC 118
4.16 ZrC–TaC 119
4.17 ZrB$_2$–SiC 119
4.18 ZrB$_2$–ZrN$_{0.96}$ 121
4.19 ZrB$_2$–W$_2$B$_5$ 121
4.20 Hf–Zr–C (Hf, Zr)C$_{1-x}$ ss) 122
4.21 HfC–ZrC 124
4.22 HfC–TaC 124
4.23 HfB$_2$–HfC$_{0.9}$ and HfB$_2$–B$_4$C 125
4.24 HfB$_2$–GdB$_6$ 126
4.25 HfB$_2$–W$_2$B$_5$ 129
4.26 HfB$_2$–MoB$_2$ 130

4.27   TiN–AlN   130
4.28   ZrN–AlN   131
4.29   ZrN–ZrO$_2$–Y$_2$O$_3$.   132
4.30   NbC–VC–Ni   132
4.31   NbC–VC   134
4.32   NbB$_2$–B$_4$C   134
4.33   TaB$_2$–B$_4$C   135
4.34   VB$_2$–B$_4$C   135
4.35   VB$_2$–VC$_{0.88}$ 137
4.36   W$_2$B$_5$–B$_4$C, W$_2$B–W$_2$C, WB–W$_2$C, etc.   137
4.37   TiC–HfC–VC   139
4.38   TiC–HfC–WC   141
4.39   TiC–HfC–(MoC)   142
4.40   TiC–WC–NbC   145
4.41   TiC–WC–TaC   146
4.42   TiC–TaC/NbC–WC   146
4.43   HfC–NbC–VC   147
4.44   HfC–VC–WC   148
4.45   HfC–VC–(MoC)   151
4.46   NbC–TaC–WC   151
4.47   V$_2$C–Ta$_2$C–W$_2$C   153
4.48   V$_2$C–Ta$_2$C–Mo$_2$C   154
4.49   Nb$_2$C–Ta$_2$C–W$_2$C   155
4.50   Nb$_2$C–Ta$_2$C–Mo$_2$C   156
4.51   ZrC–ZrO$_2$–SiC–SiO$_2$–MgO   157
4.52   ZrC–ZrO$_2$–SiC–SiO$_2$–CaO   159
       References   159

**Erratum to: Phase Equilibria Diagrams of High Temperature
Non-oxide Ceramics** E1

Appendices.   163

Chapter 1
Si_3N_4 Ceramics Systems

Abstract Phase diagrams of Si_3N_4-based ceramics systems which include two subsystems of Si_3N_4–M_xX_y and various SiAlON solid solutions are collected in this chapter. M_xX_y are oxides, rare earth oxides, or certain non-oxide compound usually as sintering aids of Si_3N_4 ceramic. At high temperature, M_xX_y react with Si_3N_4 forming diverse nitrogen-containing silicates and aluminosilicates reserved as secondary phase in these systems. Phase diagrams of SiAlON systems can be a guide for manufacturing of the SiAlONs ceramics. Five different types of SiAlON ceramics with significance in crystal chemistry and physical chemistry of nitrogen ceramics are presented in this chapter, namely, (a) β–SiAlON (β–Si_3N_4 ss), (b) α–SiAlON (α–Si_3N_4 ss), (c) O′SiAlON (Si_2ON_2–Al_2O_3 ss), (d) M′(R)–SiAlON ("$Si_3N_4 \cdot R_2O_3$"–Al_2O_3 ss), and (e) SiAlON polytypoids.

1.1 Si_3N_4–M_xX_y Systems

Back in the 1970s, silicon nitride as an excellent structural ceramic was thought of potential for application in turbo cars. For this particular reason, physical chemistry and phase equilibrium diagrams of the Si_3N_4–M_xX_y (M = certain metals) systems were extensively studied, and thus initiated lasting researches on the whole field of nitride ceramics, including the Si_3N_4-based nitrogen-containing silicates and the M–Al–Si oxynitrides (see Appendix Tables A.1–A.4). Structural units of Si_3N_4 and Si_2ON_2 are [SiN_4] and [$SiON_3$] tetrahedron (see Appendix Table A.5), respectively. Because these groups are acidic same as [SiO_4] in SiO_2, they react with basic oxides of metals or rare earths to form a series of refractory nitrogen-containing silicates. Therefore, oxides of metals or rare earths are usually used as sintering aids to the nitride ceramics. Many of such nitrogen-containing silicates are closely related with traditional silicates or minerals where found many homotypes (see Appendix Tables A.3 and A.4). Meanwhile, discovery of SiAlONs as solid solutions of silicon nitride significantly motivated researches on the nitride ceramics.

© Springer Nature Singapore Pte Ltd. 2018
Z. Huang and L. Wu, *Phase Equilibria Diagrams of High Temperature Non-oxide Ceramics*, https://doi.org/10.1007/978-981-13-0463-7_1

Knowledge advances surely would help optimize compositional design and processing of the nitride ceramics. This section will introduce the high temperature reactions and phase equilibrium diagrams of the Si$_3$N$_4$–M$_x$X$_y$ systems.

1.1.1   Si$_3$N$_4$–SiO$_2$

Figure 1.1 shows the calculated phase diagrams of SiO$_2$–Si$_3$N$_4$ system, where Fig. 1.1a is picked up from a large program for calculating phase diagrams of the Si–C–N–O system by the Powder-Metallurgic Laboratory (PML), Max Planck

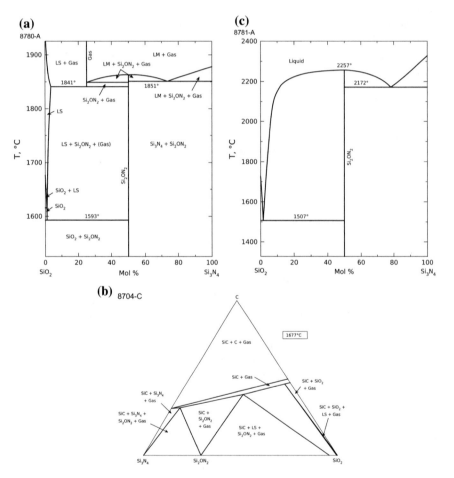

Fig. 1.1 a Calculated phase diagram of SiO$_2$–Si$_3$N$_4$ system. *LM* liquid metal, *LS* liquid salt. Reprinted with permission of The American Ceramic Society, www.ceramics.org. **b** Calculated isothermal section of Si$_3$N$_4$–C–SiO$_2$ system at 1677 °C. Reprinted with permission of The American Ceramic Society. **c** Calculated phase diagram of SiO$_2$–Si$_3$N$_4$ system. Reprinted with permission of The American Ceramic Society

Institute, Germany, originally published in [1]. PML calculated the phase diagrams by the "Regular Solution Model" and using the JANAF thermodynamic database [2]. Data for Si and Si_2ON_2 were taken from Refs. [3, 4], respectively. Elaboration on the calculation can be found in Ref. [1] and Figs. 8745, 8703, and 8704 in ACerS-NIST Phase Equilibria Diagrams PC Database Version 4.0 (PED).

Figure 1.1b shows the isothermal section of Si_3N_4–C–SiO_2 system at 1677 °C [1]. Figure 1.1c shows the phase diagram of SiO_2–Si_3N_4 system [5] calculated by "Regular Solution Model", without considering the known decomposition of $Si_3N_4(s) \rightarrow Si(l) + N_2(g)$ and $Si_2ON_2(s) \rightarrow Si(l) + N_2(g) + SiO(g)$. Method for calculating the SiO_2–Si_3N_4 phase diagram was described in PED 8785.

1.1.2 Si_3N_4–Y_2O_3

Figure 1.2 shows the pseudobinary diagram of Si_3N_4–Y_2O_3 system [6]. Samples were sintered under 1 MPa N_2 to prevent compositional volatilization. X-ray diffraction (XRD) analysis showed two compounds of $Y_2Si_3O_3N_4$ (the 1:1 compound, tetragonal M-phase, N-melilite) and $2Y_2O_3Si_2ON_2$ (monoclinic J-phase, cuspidine-type). The presence of J-phase in the $3Y_2O_3$:Si_3N_4 (3:1) sample was due to oxygen impurity in the Si_3N_4 starting powder participating in reaction. Although J-phase can be coexistent with both M-phase and Y_2O_3, its composition not lies in the Si_3N_4–Y_2O_3 binary system. Therefore, the J-phase is presented by dotted line in Fig. 1.2. The $85Y_2O_3$:$15Si_3N_4$ (mol%) composition on the M–Y_2O_3 tie line has a

Fig. 1.2 Pseudobinary diagram of Si_3N_4–Y_2O_3 system. Reprinted from Ref. [6], Copyright 1996, with permission from John Wiley and Sons

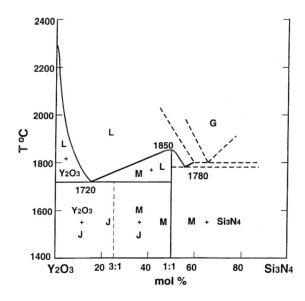

eutectic temperature (T$_{eu}$) of 1720 °C, actually the melting temperature of the J-phase. Compositions in Si$_3$N$_4$–rich area have significant weight losses at 1780 °C, indicating release of gaseous phases. For these reasons, eutectic points and liquidus lines could not be set unanimously; therefore, they are presented by dotted lines to show a pseudobinary system.

1.1.3   Si$_3$N$_4$–Y$_3$Al$_5$O$_{12}$

Figure 1.3 shows a phase diagram of Si$_3$N$_4$–Y$_3$Al$_5$O$_{12}$ binary system. Samples were fired under 1 MPa N$_2$ to prevent compositional volatilization. However, a significant weight loss was observed at temperatures above 1700 °C. The eutectic composition at T$_{eu}$ = 1650 °C is 25 Si$_3$N$_4$ + 75 YAG (Y$_3$Al$_5$O$_{12}$) by weight [7].

In order to prevent compositional volatilization in Si$_3$N$_4$ ceramic system, higher nitrogen pressure or hot pressing (HP) would be better to use.

1.1.4   Si$_3$N$_4$–Y$_2$Si$_2$O$_7$

Figure 1.4 shows a phase diagram of Si$_3$N$_4$–Y$_2$Si$_2$O$_7$ binary system. Experimental conditions were the same as that for Figs. 1.2 and 1.3. The eutectic composition at T$_{eu}$ = 1570 °C is 10Si$_3$N$_4$ + 90 Y$_2$Si$_2$O$_7$ by weight [7].

Fig. 1.3 Phase diagram of Si$_3$N$_4$–Y$_3$Al$_5$O$_{12}$ binary system. Reprinted, from Ref. [7], Copyright 1994, with permission from John Wiley and Sons

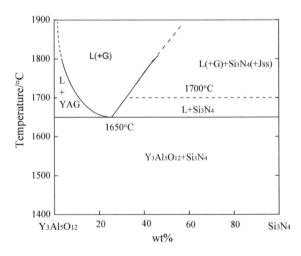

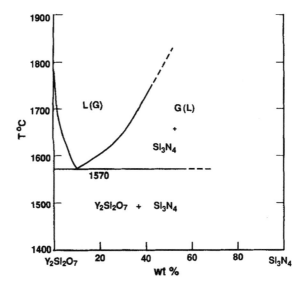

Fig. 1.4 Phase diagram of Si₃N₄–Y₂Si₂O₇ system. Reprinted, from Ref. [7], Copyright 1994, with permission from John Wiley and Sons

1.1.5 *Si₃N₄–Y₃Al₅O₁₂–Y₂Si₂O₇*

Figure 1.5 shows the isothermal section of Si₃N₄–Y₃Al₅O₁₂–Y₂Si₂O₇ system at 1650 °C under 1 MPa N₂. There exists a liquid area in the oxide-rich side. The eutectic composition was detected as 10 Si₃N₄ + 27 YAG + 63 Y₂Si₂O₇ by weight at T_{eu} = 1430 °C. Samples with compositions in the liquid area can result in homogenously transparent nitrogen-containing glass [7].

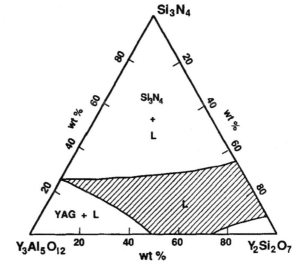

Fig. 1.5 Isothermal section of Si₃N₄–Y₃Al₅O₁₂–Y₂Si₂O₇ ternary system at 1650 °C. Reprinted from Ref. [7], Copyright 1994, with permission from John Wiley and Sons

1.1.6   Si$_3$N$_4$–SiO$_2$–La$_2$O$_3$

Figure 1.6a shows the isothermal section of Si$_3$N$_4$–SiO$_2$–La$_2$O$_3$ system at 1700 °C [8]. Samples were prepared by hot pressing at 1400 °C under 20 MPa or by pressureless sintering at 1700 °C. Electron diffraction, XRD, and electronic probe were used to determine chemistries of the phases in equilibrium. There are two new compounds which were identified as La$_2$Si$_6$O$_3$N$_8$ (1:2, monoclinic structure) and LaSiO$_2$N (K-phase, hexagonal structure). LaSiO$_2$N and another phase of La$_4$Si$_2$O$_7$N$_2$ (J-phase monoclinic) could decompose to result in La$_5$(SiO$_4$)$_3$N (H-phase, hexagonal La-apatite) and a glassy phase. In addition, there exists an undetermined liquid area, from which K- and J-phase could precipitate.

Fig. 1.6 a Isothermal section of Si$_3$N$_4$–SiO$_2$–La$_2$O$_3$ system at 1700 °C. Reprinted from Ref. [8], Copyright 1982, with permission from Springer Nature. **b** Subsolidus phase diagram of Si$_3$N$_4$–SiO$_2$–La$_2$O$_3$ system. Reprinted from Ref. [9], Copyright 2011, with permission from John Wiley and Sons

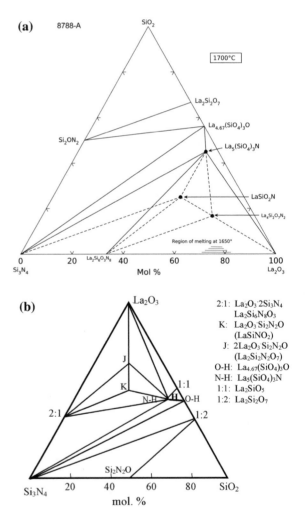

Figure 1.6b shows the subsolidus phase diagram of Si_3N_4–SiO_2–La_2O_3 system [9]. Four compounds, i.e., $La_2O_3{\cdot}2Si_3N_4$ (monoclinic), $2La_2O_3{\cdot}Si_2N_2O$ (J-phase, Cuspidine-type), $La_2O_3{\cdot}Si_2N_2O$ (K-phase, Wollastonite-type), and $La_5(SiO_4)_3N$ (NH-phase, Nitrogen-containing La-Apatite) exist in the diagram. They were easy to form by solid-state reaction under hot pressing at 1600 °C. Formation of the apatite phase needed more SiO_2 to be incorporated in the apatite lattice. It was easier for La_2O_3 with stronger ionicity relative to other rare earth oxides such as Gd_2O_3 and Y_2O_3 (ionicity follows the order of $La_2O_3 > Gd_2O_3 > Y_2O_3$) to react with the covalent Si_3N_4, leaving more SiO_2 for H(La)-apatite to form. Besides, a small amount of solid solution between NH and OH ($La_{4.67}(SiO_4)_3O$, oxygen-containing La-apatite) also forms in the system.

1.1.7 Si_3N_4–SiO_2–Ce_2O_3

Ceria CeO_2 with Ce^{4+} is often used as starting powder, because it is more stable and easier to be obtained. Ce_2O_3 with Ce^{3+} is resulted from reactions with Si_3N_4 at high temperatures:

$$2Si_3N_4 + 6CeO_2 \rightarrow 3Si_2N_2O + 3Ce_2O_3 + N_2$$

$$Si_3N_4 + 12CeO_2 \rightarrow 3SiO_2 + 6Ce_2O_3 + 2N_2$$

Subsolidus phase diagram of the Si_3N_4–SiO_2–Ce_2O_3 system at 1650–1750 °C [10] was proposed, as shown in Fig. 1.7. Samples were prepared by hot pressing or

Fig. 1.7 Tentative phase diagram of Si_3N_4–SiO_2–Ce_2O_3 system at 1650–1750 °C. Reprinted with permission of The American Ceramic Society

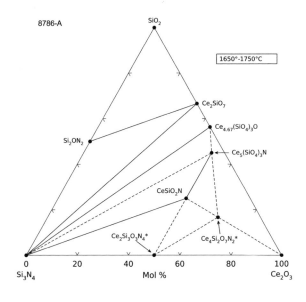

pressureless sintering process. K-phase was experimentally obtained. Three other ternary phases, namely Ce$_5$(SiO$_4$)$_3$N, Ce$_2$Si$_3$O$_3$N$_4$, and Ce$_4$Si$_2$O$_7$N$_2$, were assumedly compatible with Si$_3$N$_4$, in analogy to the Y-system (see PED 8790). The asterisks indicate phases in expect but were not experimentally observed yet. Dashed lines are tentative or hypothetical.

1.1.8   Si$_3$N$_4$–SiO$_2$–Gd$_2$O$_3$

Figure 1.8 shows the subsolidus phase diagram of Si$_3$N$_4$–SiO$_2$–Gd$_2$O$_3$ system [9]. Except for the absence of K-phase (GdSiO$_2$N), the existence of other phases and the phase relations are the same as that in the Si$_3$N$_4$–SiO$_2$–Y$_2$O$_3$ system (see PED 8789, 8790).

1.1.9   Si$_3$N$_4$–SiO$_2$–Y$_2$O$_3$

Besides the yttrium silicates in the SiO$_2$–Y$_2$O$_3$ subsystem, the formation of four ternary nitrogen-containing silicates, i.e., Y$_{10}$(SiO$_4$)$_6$N$_2$ (hexagonal N-apatite), YSiO$_2$N (hexagonal N-Wollastonite), Y$_4$Si$_2$O$_7$N$_2$ (monoclinic cuspidine), and Y$_2$Si$_3$O$_3$N$_4$ (tetragonal N-melilite), was evidenced in the Si$_3$N$_4$–SiO$_2$–Y$_2$O$_3$ system. There are inconsistencies in open literatures [9, 11–13] about the subsolidus phase relations in the Si$_3$N$_4$–SiO$_2$–Y$_2$O$_3$ system in Fig. 1.9a–d, coming from the following questions: (1) Which tie line is more suitable between Si$_3$N$_4$–Y$_2$Si$_2$O$_7$ and Si$_2$N$_2$O–H(Y)? and (2) Does the Y$_2$O$_3$–H(Y) tie line exist or not? The same controversy is also involved in the Si$_3$N$_4$–SiO$_2$–La$_2$O$_3$ system due to similarity between the rare earths elements. Single-phase Si$_2$N$_2$O or coexistence of Si$_2$N$_2$O with any

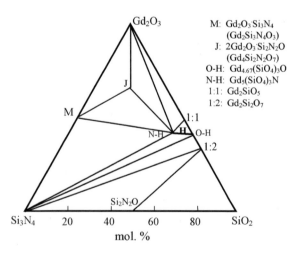

Fig. 1.8 Subsolidus phase diagram of Si$_3$N$_4$–SiO$_2$–Gd$_2$O$_3$ system. Reprinted from Ref. [9], Copyright 2011, with permission from John Wiley and Sons

M: Gd$_2$O$_3$ Si$_3$N$_4$ (Gd$_2$Si$_3$N$_4$O$_3$)
J: 2Gd$_2$O$_3$ Si$_2$N$_2$O (Gd$_4$Si$_2$N$_2$O$_7$)
O-H: Gd$_{4.67}$(SiO$_4$)$_3$O
N-H: Gd$_5$(SiO$_4$)$_3$N
1:1: Gd$_2$SiO$_5$
1:2: Gd$_2$Si$_2$O$_7$

R-siliconoxynitride was not detected in all the hot-pressed samples investigated in
[9], though R-siliconoxynitrides are formed on the R_2O_3–Si_2N_2O tie line. It means
that Si_2N_2O and R-siliconoxynitride are not compatible on the R_2O_3–Si_2N_2O
(R = La, Gd, Y) tie line. Instead, reaction of $R_2O_3 + 4Si_2N_2O = 2Si_3N_4 + R_2Si_2O_7$
could be true. Therefore, Si_2N_2O–H should be replaced by Si_3N_4–$R_2Si_2O_7$ tie line.

With regard to the R_2O_3–H, this tie line could be joined because of compatibility
of R_2O_3–H–J, indicating the Y_2O_3–H tie line in Fig. 1.9c be more reasonable than
the Y_2SiO_5–J tie line as shown in Fig. 1.9b.

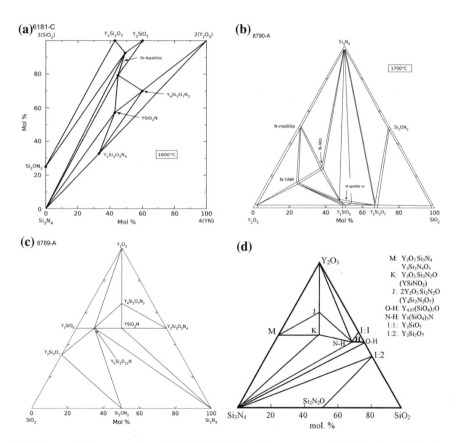

Fig. 1.9 a Isothermal section of Si_3N_4–YN–Y_2O_3–SiO_2 system at 1600 °C. Reprinted with
permission of The American Ceramic Society. **b** "Behavior" diagram of Si_3N_4–Y_2O_3–SiO_2 system
at 1700 °C. Reprinted with permission of The American Ceramic Society. **c** Subsolidus phase
diagram of Si_3N_4–SiO_2–Y_2O_3 system. Reprinted with permission of The American Ceramic
Society. **d** Subsolidus phase diagram of Si_3N_4–SiO_2–Y_2O_3 system. Reprinted from Ref. [9]
Copyright 2011, with permission from John Wiley and Sons

1.1.10 Si$_3$N$_4$–La$_2$O$_3$–Y$_2$O$_3$

Samples were hot pressed at 1550–1800 °C for 1 h in N$_2$ atmosphere. A new phase with composition of 1.7Y$_2$O$_3$·3.3La$_2$O$_3$·15Si$_3$N$_4$ (Y$_{3.4}$La$_{6.6}$Si$_{45}$O$_{15}$N$_{60}$) was found by XRD analysis near the Si$_3$N$_4$-rich region [14]. Solid solubility of La$_2$O$_3$:Si$_3$N$_4$ (1:1) in Y$_2$Si$_3$O$_3$N$_4$ (1:1) kept unchanged at 1550–1800 °C (see Fig. 1.10a). Continuous solid solutions were formed on La$_2$Si$_2$O$_4$N$_2$–Y$_2$Si$_2$O$_4$N$_2$ (K-phase), La$_4$Si$_2$O$_7$N$_2$–Y$_4$Si$_2$O$_7$N$_2$ (J-phase), and La$_5$Si$_3$O$_{12}$N–Y$_5$Si$_3$O$_{12}$N (H-phase) tie line, respectively (see Fig. 1.10b).

Fig. 1.10 a Subsolidus phase diagram of Si$_3$N$_4$–La$_2$O$_3$–Y$_2$O$_3$ ternary system. Reprinted with permission of The American Ceramic Society. **b** Tri-prism diagram of Si$_3$N$_4$–LaN–YN–La$_2$O$_3$–Y$_2$O$_3$–SiO$_2$ hexahydric system. Reprinted with permission of The American Ceramic Society

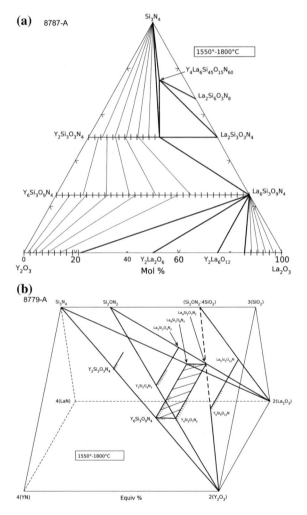

1.1.11 *Si₃N₄–SiO₂–Th₃N₄–ThO₂*

Figure 1.11 shows the subsolidus phase diagram of Si$_3$N$_4$–SiO$_2$–Th$_3$N$_4$–ThO$_2$ system [15]. It is consisted of tie lines, especially tie lines showing coexistence of ThO$_2$ with other compounds. Neither ternary nor quaternary compound formed.

1.1.12 *Si₃N₄–SiO₂–AlN–Al₂O₃–ZrN–ZrO₂*

Because of the importance of reactions taking place at high temperatures for sintering of silicon nitride ceramics, this system had been paid much more attention. Figure 1.12a shows the "behavior" diagram of Si$_3$N$_4$–SiO$_2$–Zr$_3$N$_4$–ZrO$_2$ system at 1700 °C [12], the importance of which lies in the various oxides used as additives necessary for liquid phase sintering of silicon nitride ceramics. Pellets with composition of Si$_3$N$_4$ + ZrO$_2$ or β–SiAlON (Si$_3$Al$_3$O$_3$N$_5$) + ZrO$_2$ were compacted by cold pressing in a steel die, followed by dividing each pellet into two pieces, one piece for hot pressing and the other for pressureless sintering. Hot pressing was performed under a mechanical pressure of 31.7 MN/m^2 at 1800 °C. For pressureless sintering, the pellets were embedded in BN powder inside an alumina crucible and were heated in a slow stream of N$_2$ at atmospheric pressure.

The term "behavior" diagram was used to emphasize that it was not proven whether equilibrium had been reached or not. ZrN could be produced when

Fig. 1.11 Subsolidus phase diagram of Si$_3$N$_4$–SiO$_2$–Th$_3$N$_4$–ThO$_2$ system. Reprinted with permission of The American Ceramic Society

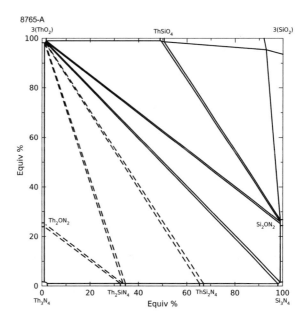

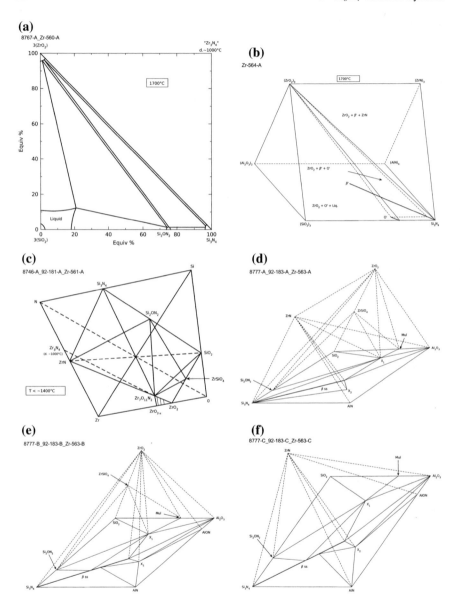

Fig. 1.12 a "Behavior" diagram of Si$_3$N$_4$–SiO$_2$–Zr$_3$N$_4$–ZrO$_2$ system at 1700 °C. Reprinted with permission of The American Ceramic Society. **b** "Behavior" diagram of Si$_3$N$_4$–AlN–ZrN–SiO$_2$–Al$_2$O$_3$–ZrO$_2$-system at 1700 °C. Reprinted with permission of The American Ceramic Society. **c** Phase diagram of N–Si–O–Zr system. Reprinted with permission of The American Ceramic Society. **d** Prism diagram of ZrO$_2$–ZrN–Si$_3$N$_4$–AlN–Al$_2$O$_3$–SiO$_2$ hexahydric system. Reprinted with permission of The American Ceramic Society. **e** Subsolidus phase diagram of ZrO$_2$–Si$_3$N$_4$–AlN–Al$_2$O$_3$–SiO$_2$ quinary subsystem. Reprinted with permission of The American Ceramic Society. **f** Subsolidus phase diagram of ZrN–Si$_3$N$_4$–AlN–Al$_2$O$_3$–SiO$_2$ quinary subsystem. Reprinted with permission of The American Ceramic Society

compositions were located in the Si_3N_4–Si_2ON_2–ZrO_2 triangle if weight loss took place due to volatile SiO and N_2. See PED 8746 for this system at lower temperatures.

Figure 1.12b shows the "behavior" diagram of $(ZrO_2)_3$–$(Al_2O_3)_2$–$(SiO_2)_3$–Si_3N_4–$(AlN)_4$–$(ZrN)_3$ system [16], in which $\beta' = Si_{6-z}Al_zO_zN_{8-z}$ and $O' = Si_{2-x}Al_xO_{1+x}N_{2-x}$.

The authors [16] studied a group of compositions with varying O':β' ratios and with different sintering additives in order to examine phase equilibria in the ZrO_2–O'–β' region and densification behavior of the materials. It was expected to improve toughness of the ceramics by adding β' to ZrO_2/O' composites. However, increasing in nitrogen content results in increasing of the ZrN content. ZrN hindered toughness and degraded oxidation resistance at temperatures above 600 °C.

Starting powders were commercially available Si_3N_4, Al_2O_3, monoclinic ZrO_2, and $ZrSiO_4$. The ZrO_2 amount was fixed at 25 wt% in all samples and the ratio of O':β' was varied from about 1.5 to zero ($O' = Si_{1.85}Al_{0.15}O_{1.15}N_{1.85}$, $\beta' = Si_{5.2}Al_{0.8}O_{0.8}N_{7.2}$). Sm_2O_3 and Y_2O_3 were added to some compositions as sintering aids. Samples were sintered in a carbon resistance furnace at 1600, 1650, and 1700 °C for 1 h. Reaction was complete at 1700 °C. Specimens were examined by X-ray diffraction.

Formation of ZrN was favored as the composition shifted from the O' to the β' region. ZrN was always in equilibrium with pure β' at 1650–1700 °C. If O' was present in the product, ZrN was eliminated. The ZrO_2–β'–O' compatibility volume as shown in the diagram indicated a region in which these composites were formed without ZrN. See PED Zr-564 for this system.

Figure 1.12c shows schematic phase relations and compatibility of principal oxides, nitrides, and oxynitrides in the N–Zr–O–Si system at $T < \sim 1400$ °C [17].

Compositions along the tie lines indicated were prepared at $1400° < T < 1800$ °C. Starting materials were Si_3N_4, SiO_2, ZrN, and ZrO_2 of variable purity, generally above 98%. Specimens were either hot pressed under flowing N_2 in BN-coated graphite dies or pressureless sintered under N_2 as prepressed (35 MPa) pellets packed in BN powder. Two major reactions involving gaseous phases were noted in the phase diagram and are shown by dashed lines, indicating inadequacy of condensed phase representation.

$$4Si_3N_4 + 6ZrO_2 = 12SiO(g) + 6ZrN + 5N_2(g) \quad (T > 1600 \,°C) \tag{1}$$

$$2SiO_2 + ZrN = ZrO_2 + 2SiO(g) + 0.5N_2(g) \quad (T > 1400 \,°C) \tag{2}$$

Oxygen partial pressure had a significant influence on equilibria of the condensed phase, especially affecting stability of Zr-containing phases. At low P_{O^2} values, the nitride phases were stable with Zr^{3+} valence, while at high P_{O^2}, oxygen-rich phases with Zr^{4+} were formed. Stabilization of cubic ZrO_2 by nitrogen was also noted by the authors [17] but the complete solubility range has not been determined.

Figure 1.12d–f shows the phase relations in ZrO_2–ZrN–Si_3N_4–AlN–Al_2O_3–SiO_2 system at temperatures ranging from 1525–1625 °C [17]. Starting powders were Si_3N_4, SiO_2, ZrN, ZrO_2, and AlN. Samples were hot pressed or pressureless sintered at 1400 °C $< T <$ 1800 °C. The research resulted in a prism diagram as shown in Fig. 1.12d. It is consisted of two quinary systems, Fig. 1.12e, f, in which ZrO_2 and ZrN, respectively, coexisted with other phases on the SiAlON plane. The cation valence of Zr in the products was related to oxygen partial pressure. At low P_{O^2} values, the nitride phases having Zr^{3+} were stable, while at high P_{O^2} the oxygen-rich phases with Zr^{4+} were formed, indicating the happening of reaction $8AlN + 6ZrO_2 = 6ZrN + 4Al_2O_3 + N_2(g)$ ($T \sim$ 1700 °C).

1.1.13 Si_3N_4–SiO_2–ZrN–ZrO_2–Y_2O_3

Starting powders were Si_3N_4, ZrN, and Y_2O_3. Mixtures of the starting powders in proper proportions were blended by hand with a pestle and mortar for 2 h using absolute alcohol as medium. After being dried, the powder mixtures were hot pressed in nitrogen atmosphere at 1750 °C for 1 h under 16 $\sim$ 31 MPa pressure. Chemical composition of each phase was analyzed by energy-dispersive X-ray analysis (EDX). Three nitrogen-containing yttrium silicates, i.e., $Y_2Si_3O_3N_4$ (M-phase, Y-melilite), $Y_4Si_2O_7N_2$ (J-phase, Y-cuspidine), and $Y_5 (SiO_4)_3N$ (H-phase, Y-apatite), had formed. ZrN was detected, indicating no reaction happens between ZrN and the end members in the Si_3N_4–ZrN–Y_2O_3 system. Formation of the two oxygen-rich phases, i.e., J- and H-phases, was due to SiO_2 impurity in the Si_3N_4 powder. Therefore, SiO_2 was taken into the system for investigation. A quaternary system of Si_3N_4–SiO_2–ZrN–Y_2O_3 was also studied. As a result, ZrN was evidenced in equilibrium with each of the silicates, as shown in Fig. 1.13a, b [18]. In addition, phase relations in the ZrN–ZrO_2–Y_2O_3 system were also deter-mined, which showed the coexistence of ZrN with Css, Fss, and T–ZrO_2 solid solutions in the ZrO_2–Y_2O_3 subsystem. Based on these results, the authors pre-sented the subsolidus phase diagram of the ZrN–ZrO_2–Si_3N_4–SiO_2–Y_2O_3 quinary system, as shown in Fig. 1.13c [18].

1.1.14 Si_3N_4–SiO_2–Mg_3N_2–MgO

Some phase diagrams of this system have been reported, with two different joins being presented by different researchers [12, 19–23] in Fig. 1.14a–c. Figure 1.14a shows the "Behavior" diagram of this (Si–Mg–O–N) system at 1700 °C [12]; the dashed line shows a liquid region at 1700 °C, and the homogeneous glass com-positions are denoted by G found on cooling; (b) shows the solid phase relations of the system at $T < \approx$1400 °C [19]; (c) superposes work of different authors, in

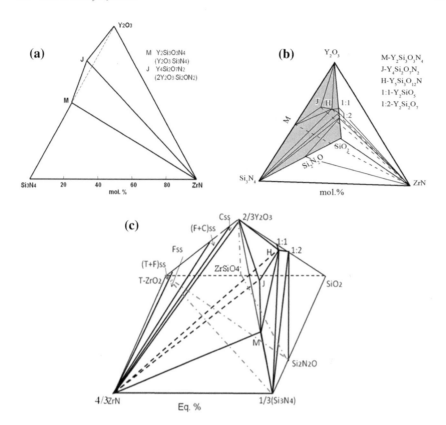

Fig. 1.13 **a** Subsolidus phase diagram of Si₃N₄–ZrN–Y₂O₃ system. Reprinted from Ref. [18], Copyright 2015, with permission from Journal of the Chinese Ceramic Society. **b** Subsolidus phase diagram of Si₃N₄–SiO₂–ZrN–Y₂O₃ system. Reprinted from Ref. [18], Copyright 2015, with permission from Journal of the Chinese Ceramic Society. **c** Tentative diagram of Si₃N₄–SiO₂–ZrN–ZrO₂–Y₂O₃ system

which solid lines are from reference [21], dotted lines are from [22] and PED 8762, and dot dashed lines are from [23].

A review paper [20] compared the results by three laboratories. The join Mg₂SiO₄–MgSiN₂ was reported [21, 22], but the binary join Si₃N₄–MgO was formed when samples were hot pressed at 1750 °C [23]. An independent study [20] showed that both Si₃N₄–MgO and Mg₂SiO₄–MgSiN₂ joins can be seen depending on the firing conditions being used. The samples fired at 1550 °C had lower weight loss, and the former join was observed without reaction; with higher weight loss, the latter join was observed, i.e., the reaction Si₃N₄ + 4MgO → Mg₂SiO₄ + 2MgSiN₂ happened.

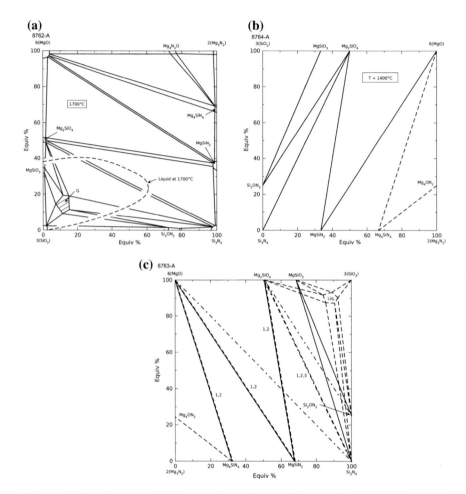

Fig. 1.14 a "Behavior" diagram of 6(MgO)–3(SiO$_2$)–Si$_3$N$_4$–2(Mg$_3$N$_2$) system at 1700 °C. Reprinted with permission of The American Ceramic Society. **b** Solid phase diagram of 3(SiO$_2$)–Si$_3$N$_4$–2(Mg$_3$N$_2$)–6(MgO) quasi-quaternary system at $T \approx$ 1400 °C. Reprinted with permission of The American Ceramic Society. **c** Phase diagram of 6(MgO)–2(Mg$_3$N$_2$)–Si$_3$N$_4$–3(SiO$_2$) system. Reprinted with permission of The American Ceramic Society

1.1.15 Si$_3$N$_4$–Si$_2$ON$_2$–Mg$_2$SiO$_4$

Figure 1.15 shows a calculated pseudoternary phase diagram of Si$_3$N$_4$–Si$_2$ON$_2$–Mg$_2$SiO$_4$ system by using composition members of (a) Si$_3$N$_4$, Si$_2$ON$_2$ and Mg$_2$SiO$_4$, and (b) Si$_3$N$_4$, MgO, and SiO$_2$ [5].

The calculation employed a simple regular solution model with nine regression (or "excess") parameters for binary and ternary interactions in each case. For each diagram, the calculated result was the best fit with the experimental data.

Fig. 1.15 a Calculated by using composition members of Si$_3$N$_4$, Si$_2$ON$_2$, and Mg$_2$SiO$_4$. Reprinted with permission of The American Ceramic Society. **b** Calculated by using Si$_3$N$_4$, MgO, and SiO$_2$ compositions. Reprinted with permission of The American Ceramic Society

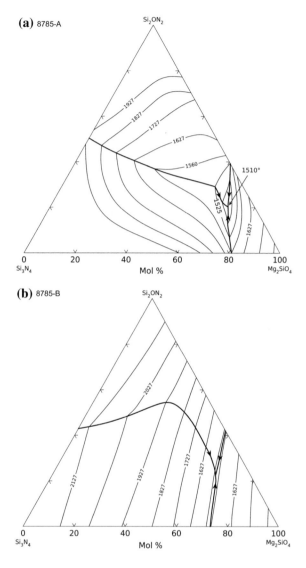

The system described in Fig. 1.15(a) is pseudoternary as it does not consider the known decomposition of the compounds Si$_3$N$_4$(s) and Si$_2$ON$_2$(s) to Si(l) + N$_2$(g) plus SiO(g) (in the case of the latter compound). In that calculation, the eutectic was found to appear at 1510 °C with a composition of 0.153 Si$_3$N$_4$ + 0.194 Si$_2$ON$_2$ + 0.653 Mg$_2$SiO$_4$.

It is interesting to note that the calculated results of Fig. 1.15a are consistent with the results of the system Si$_3$N$_4$–Si$_2$ON$_2$ in which the metallic Si liquid and gas phases were considered explicitly. Figure 1.15b is in better agreement with calculated condensed phase equilibria on this eutectic location [5].

1.1.16   Si$_3$N$_4$–Ca$_3$N$_2$–Mg$_3$N$_2$–MgO–SiO$_2$ (+CaO)

Figure 1.16a shows the partial hypothetical subsolidus compatibility of Ca$_3$N$_2$–Mg$_3$N$_2$–MgO–SiO$_2$–Si$_3$N$_4$ system. The dashed line illustrates the join as shown in Fig. 16b [24].

Figure 1.16b shows the join between the ternary eutectic (0.11Si$_3$N$_4$ + 0.34SiO$_2$ + 0.55MgO) and its CaO counterpart [24].

Starting materials were Si$_3$N$_4$, SiO$_2$, MgO, and CaO powders of unspecified source or purity, which were milled with WC, pressed, and hot pressed in graphite dies in N$_2$ under uniaxial stress of 7 MPa for 30 min. Temperatures were kept

Fig. 1.16 a Partial hypothetical subsolidus compatibility of Ca$_3$N$_2$–Mg$_3$N$_2$–MgO–SiO$_2$–Si$_3$N$_4$ system. Reprinted with permission of The American Ceramic Society. **b** The join between the ternary eutectic composition and its CaO counterpart. Reprinted with permission of The American Ceramic Society

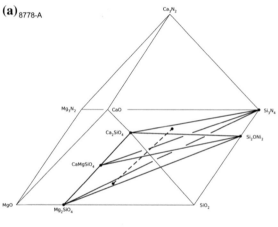

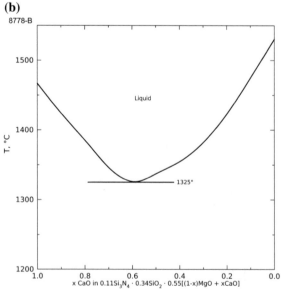

below 1600 °C for determining the lowest melting composition, the eutectic. Three eutectics were found: a binary at 7 mol% Si$_3$N$_4$-93 mol% Mg$_2$SiO$_4$ at 1560 °C; another binary at 15 mol% Si$_2$ON$_2$-85 mol% Mg$_2$SiO$_4$ at 1525 °C; and a ternary at 4 mol% Si$_3$N$_4$-14 mol% Si$_2$ON$_2$-82 mol% Mg$_2$SiO$_4$ at 1515 °C.

1.1.17 Si$_3$N$_4$–SiO$_2$–AlN–Al$_2$O$_3$–Ca$_3$N$_2$–CaO

Figure 1.17a shows the prism diagram of Si$_3$N$_4$–4(AlN)–2(Ca$_3$N$_2$)–6(CaO)–3(SiO$_2$)–2(Al$_2$O$_3$) system and it shows the oxygen-rich region studied [25].

Figure 1.17a–e shows the diagrams showing the 14 compatibility tetrahedrals, in which CS = CaSiO$_3$, Gh = gehlenite (Ca$_2$Al$_2$SiO$_7$), An = anorthite (CaAl$_2$Si$_2$O$_8$), CA = CaAl$_2$O$_4$, Mul = mullite (Al$_6$Si$_2$O$_{13}$), X$_1$ = "nitrogen mullite", β$_{60}$ = SiAlON (Si$_{6-z}$Al$_z$O$_z$N$_{8-z}$), S = S-phase (CaO·1.33Al$_2$O$_3$·0.67Si$_2$ON$_2$), and CA$_2$ = CaAl$_4$O$_7$.

Nine phase diagrams were collected in PED 08771 (A)–(I), five of which are selected here, as shown in Fig. 1.17a–e. Twenty-nine compositions composed of the starting powders, namely Si$_3$N$_4$, AlN, CaO, Al$_2$O$_3$, and SiO$_2$, were hot pressed at different temperatures and then devitrified at 1300 °C for 20 h. In some cases, in order to verify the attainment of equilibrium state, presynthesized anorthite (CaAl$_2$Si$_2$O$_8$), gehlenite (Ca$_2$Al$_2$SiO$_7$), and CaAl$_{12}$O$_{19}$ were used as starting materials. Equilibrium was established by heating at 1550 °C for 1 h for all the compositions except the Si$_3$N$_4$-rich ones, in which transformation of the α–Si$_3$N$_4$ into β– or β′–SiAlON was very hard to complete, even at 1700 °C.

A quinary compound CaO·1.33Al$_2$O$_3$·0.67Si$_2$ON$_2$, designated as S-phase, was found. It forms continuous solid solution with CaO·2Al$_2$O$_3$. Single S-phase was obtained only by solid-state reaction.

The following compatibility tetrahedra were established in PED8771:

(A) The oxygen-rich region,
(B) SiO$_2$–An–X$_1$–Mul and An–Mul–X$_1$–Al$_2$O$_3$,
(C) SiO$_2$–CaSiO$_3$–An–Si$_2$ON$_2$ and An–X$_1$–β$_{60}$–Al$_2$O$_3$,
(D) Si$_2$ON$_2$–An–X$_1$–Si$_3$N$_4$ and Si$_3$N$_4$–X$_1$–An–β$_{60}$,
(E) Si$_2$ON$_2$–SiO$_2$–X$_1$–An and S–An–Al$_2$O$_3$–β$_{60}$,
(F) Si$_2$ON$_2$–An–Gh–Si$_3$N$_4$ and S–Gh–CaAl$_4$O$_7$–Al$_2$O$_3$,
(G) Si$_2$ON$_2$–CaSiO$_3$–An–Gh and S–An–Gh–Al$_2$O$_3$,
(H) Si$_3$N$_4$–S–An–Gh, and
 (I) Si$_3$N$_4$–β$_{60}$–S–An.

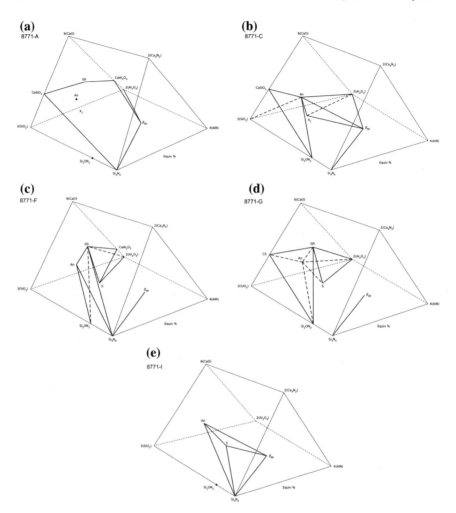

Fig. 1.17 a Prism diagram of Si₃N₄–4(AlN)–2(Ca₃N₂)–6(CaO)–3(SiO₂)–2(Al₂O₃) system. Reprinted with permission of The American Ceramic Society. **b** Compatibility tetrahedra of SiO₂–CaSiO₃–An–Si₂ON₂ and An–X1–β₆₀–Al₂O₃. Reprinted with permission of The American Ceramic Society. **c** Compatibility tetrahedra of Si₂ON₂–An–Gh–Si₃N₄ and S–Gh–CaAl₄O₇–Al₂O. Reprinted with permission of The American Ceramic Society. **d** Compatibility tetrahedra of Si₂ON₂–CaSiO₃–An–Gh and S–An–Gh–Al₂O₃. Reprinted with permission of The American Ceramic Society. **e** Compatibility tetrahedra of Si₃N₄–β₆₀–S–An. Reprinted with permission of The American Ceramic Society

1.1.18 $[Si_3N_4(60\ Mol\%)-SiO_2]-[Si_3N_4(60\ Mol\%)-Al_2O_3]-$
$[Si_3N_4(60\ Mol\%)-Y_2O_3]$

Figure 1.18 shows the isothermal section of $[Si_3N_4(60\ mol\%)-SiO_2]-$
$[Si_3N_4(60\ mol\%)-Al_2O_3]-[Si_3N_4(60\ mol\%)-Y_2O_3]$ system at 1400 °C [21], in
which SN = Si_3N_4, N = $Y_3AlSi_2O_7N_2$, Ap = apatite-type, Ak = akermanite-type,
Grt = garnet-type, X = "X-phase" or nitrogen mullite-type, Ox = Si_2ON_2,
Y = Y_2O_3, and Y$_2$S = $Y_2Si_2O_7$.

The Si_3N_4 content was held at 60 mol% for 33 compositions. Mixtures were hot
pressed at 1700 °C for 10 min in BN-coated graphite dies. The samples were then
crushed and heated at 1400 °C for 5 h in N_2 atmosphere. A new compound,
$Y_3AlSi_2O_7N_2$, was found; it is pseudohexagonal. At 1700 °C, an akermanite-type
($Y_2O_3 \cdot Si_3N_4$) or another phase ($4Y_2O_3 \cdot SiO_2 \cdot Si_3N_4$) formed solid solutions with
Al_2O_3. These solid solutions tended to decompose forming the new compound and
garnet- or apatite-type compounds at 1000–1400 °C. This system is also presented
in PED 8793. The ternary compound was not found in that study. This diagram was
published previously as Fig. 91-438.

1.2 β–SiAlON Systems

At beginning of 1970s, Oyama [26–28] and Jack [29, 30] individually discovered that
Si_3N_4 could react with Al_2O_3 at high temperature in N_2 atmosphere or in AlN presence
to form β–Si_3N_4 solid solution, in which Al^{3+} and O^{2-} substituted for Si^{4+} and N^{3-},
respectively, in Si_3N_4 lattice, but caused no change in Si_3N_4 structure [26–30].
Afterward, the fact is clear that Si_3N_4 reacts with AlN:Al_2O_3 in a same ratio of cations/

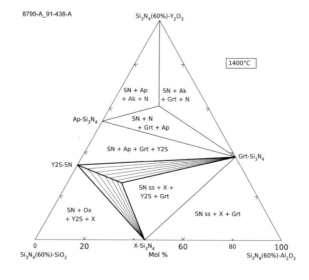

Fig. 1.18 Isothermal section
of $[Si_3N_4(60\ mol\%)-SiO_2]-$
$[Si_3N_4(60\ mol\%)-Al_2O_3]-$
$[Si_3N_4(60\ mol\%)-Y_2O_3]$
system at 1400 °C. Reprinted
with permission of The
American Ceramic Society

anions: M/X = 3/4, by displacing each other to form a substituted β–Si_3N_4 solid solution, named β–SiAlON; it refers to the four elements of Si, Al, O, and N. β–SiAlON lattice contains two molecules with β–$Si_{6-z}Al_zO_zN_{8-z}$, $0 \leq Z \leq 4.2$. Such a substitutive solid solution does not contain any vacancy or other structural defects. Its composition locates on the one-dimensional Si_3N_4–AlN:Al_2O_3 line. The bond length of Si–N in Si_3N_4 is 1.74 Å and Al–O in [AlO_4] of Al_2O_3 is 1.75 Å. Both of the nearly equal bond length and the similar hexagonal structure are the prerequisite of β–SiAlON formation (see Appendix Table A.5). Except Al–O substitutes for Si–N to form SiAlON, many small metal elements are able to enter into the Si_3N_4 lattice to form β–$(M,Al,Si)_6(N,O)_8$ solid solution, i.e., $\beta(M)$–SiAlON, M = Li, Be, Mg, Ga, Ge.

1.2.1 Si_3N_4–SiO_2–AlN–Al_2O_3 (1)

In this quaternary system, except for β–SiAlON formation, there still exist four phases of O'–SiAlON (Si_2N_2O–Al_2O_3 solid solution), AlSiONs (Si-containing AlN polytypoids), X-phase, and liquid phase. In order to clearly describe the phase relations, L. J. Gauckler et al. presented the equivalent relationship as shown in Fig. 1.19a [31]. For the different valences of Si^{+4}, Al^{+3}, N^{-3}, and O^{-2}, since 12 valences are their smallest common multiple, the four compound members also must be with 12 valences, respectively. At four middle points on the four sides of square, located these four compounds, respectively. They are all electric neutral.

Therefore, a square of Si_3N_4–Al_4N_4–Si_3O_6–Al_4O_6 quaternary system can be drawn from the four middle points with equivalent percent (see Fig. 1.19b) [31]. This two-dimensional square with equivalent percent is more distinctly showing the phase relations of the β–SiAlON systems than the three-dimensional tetrahedra using molar percentage of components.

Several experimental phase diagrams [31–34] of SiAlON systems were reported since then. They all showed that β–SiAlON is in equilibrium with O'–SiAlON, AlN polytypes, X-phase, and a liquid phase, respectively (see Fig. 1.19a–f). The X-phase of $Si_{12}Al_{18}O_{39}N_8$ has the mullite structure and a melting point of 1727 °C. The eutectic temperature T_{eu} of quaternary Si_3N_4–Al_4N_4–Si_3O_6–Al_4O_6 system is 1480 °C. Following discovery of SiAlONs, the studies of physical chemistry, equilibrium phase diagrams at high temperatures, and preparation of SiAlON ceramics all have been well developed and have been constituted a field of nitride ceramic for 40 years so far.

1.2.2 Si_3N_4–SiO_2–AlN–Al_2O_3 (2)

The section Si_3N_4–AlN–Al_2O_3–SiO_2 of the Si–Al–O–N system has been thermodynamically reassessed [35]. Improved descriptions for the Gibbs energies of the β'– and O'–SiAlON phases are applied. Different models according to the different

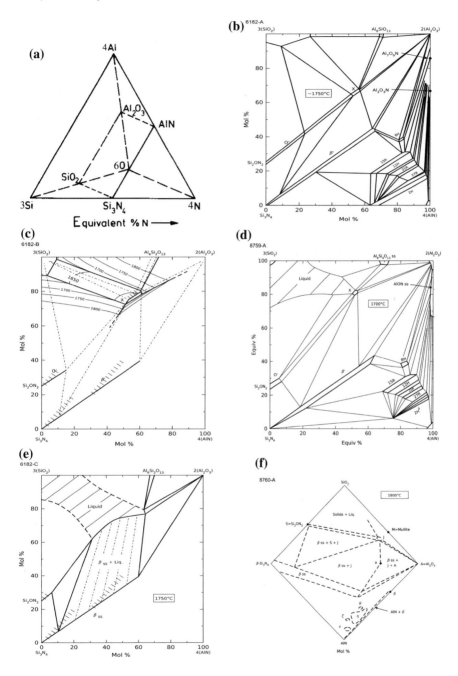

behaviors of the reciprocal system within the β–SiAlON phase are discussed. The liquid phase is modeled with a new formula based on the ionic two sublattice models. The thermodynamic properties of the SiAlON phases are discussed and

◀**Fig. 1.19** **a** SiAlON square in $3Si^{4+}$–$6O^{2-}$–$4Al^{3+}$–4 N^{3-} tetrahedral. Reprinted from Ref. [31], Copyright 1975, with permission from John Wiley and Sons. **b** Isothermal section of Si$_3$N$_4$–AlN–Al$_2$O$_3$–SiO$_2$ system at 1750 °C. Reprinted with permission of The American Ceramic Society. **c** Liquidus temperature of SiO$_2$-rich region of Si$_3$N$_4$–3SiO$_2$–4AlN–2Al$_2$O$_3$ system. Reprinted with permission of The American Ceramic Society. **d** "Behavior" diagram of Si$_3$N$_4$–3SiO$_2$–4AlN–2Al$_2$O$_3$ system at 1700 °C. Reprinted with permission of The American Ceramic Society. **e** "Behavior" diagram of Si$_3$N$_4$–4(AlN)–2(Al$_2$O$_3$)–3(SiO$_2$) system at 1750 °C. Reprinted with permission of The American Ceramic Society. **f** Quasi-equilibrium phase diagram of Si$_3$N$_4$–SiO$_2$–Al$_2$O$_3$–AlN system at 1800 °C. Reprinted with permission of The American Ceramic Society

various phase diagrams are presented. The self-consistent thermodynamic dataset is useful for the computer simulation of SiAlON synthesis. Some examples of such applications are illustrated.

Figure 1.20a shows a calculated phase diagram of Si$_3$N$_4$–AlN–Al$_2$O$_3$–SiO$_2$ system [35], in which the thick curves represent three-phase equilibria with the liquid phase; the labeled areas show the liquidus surfaces for various solids; the thin curves represent the isothermal sections.

Figure 1.20b shows the phase relations of Si$_3$N$_4$–AlN–Al$_2$O$_3$–SiO$_2$ system at 1400 °C [35];

Figure 1.20c shows the calculated maximum contents of Al in β–SiAlON at various temperatures. Dotted lines are calculated with dataset 2, solid lines with dataset 3 [35].

Figure 1.20d–f shows the selected isothermal sections of Si$_3$N$_4$–AlN–Al$_2$O$_3$–SiO$_2$ system at 1400, 1600, and 1800 °C calculated by dataset 2 [35].

It has to be suspicious of the truth that liquidus surfaces for various solids on the SiAlON plane reach such high temperatures as up to 2400–2600 °C (see Fig. 1.20a). As we know, the melting point of Si$_3$N$_4$ is 1900 °C, whereas the eutectic temperature on SiAlON plane is 1480 °C. At such high temperatures of 2400–2600 °C, SiAlON solid would have already evaporated or sublimed.

1.2.3 Si$_3$N$_4$–Li$_2$O–Al$_2$O$_3$

Figure 1.21 shows the isothermal section of Si$_3$N$_4$–Li$_2$O–Al$_2$O$_3$ pseudoternary system at 1550 °C, in which there exist many phases: spinel-type LiAl$_5$O$_8$ (M/X = 3/4), which reacts with Si$_3$N$_4$ to form β(Li)–SiAlON (Li$_{x/8}$Si$_{6-3x/4}$Al$_{5x/8}$O$_x$N$_{8-x}$); α(Li)–SiAlON (α–Si$_3$N$_4$ structure); Eu-phase LiAlSiO$_{4-x}$N$_x$ (β-eucryptite); O(Li)'–SiAlONs (Si$_2$N$_2$O-type); γ-phase tetragonal α-cristobalite (low-temperature type); 15R–SiAlON polytype (AlN structure) [36, 37]. Starting powders used were LiAlO$_2$ (melting point 1700 °C), LiAl$_5$O$_8$ (melting point 1950 °C), and Si$_3$N$_4$. Samples were hot pressed at 1550 °C under 13.8 MPa for 30 min [36, 37]. The PED' editors noted many regions in the diagram where the phase rule for a binary system was not followed; and for these regions, the diagram was designated as pseudoternary. Some difficulty in achieving equilibrium may also be inevitable in this system due to experimental difficulties.

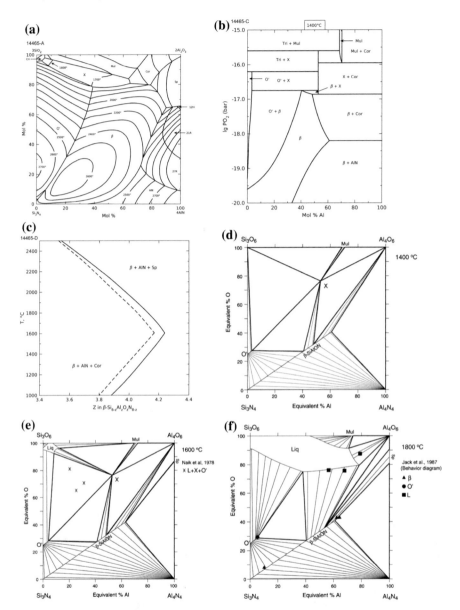

◀**Fig. 1.20** **a** Calculated phase diagram of the Si_3N_4–AlN–Al_2O_3–SiO_2 system. Reprinted with permission of The American Ceramic Society. **b** Phase relations of the Si_3N_4–AlN–Al_2O_3–SiO_2 system at 1400 °C. Reprinted with permission of The American Ceramic Society. **c** Calculated maximum contents of Al in β–SiAlON at various temperatures, dotted lines are calculated with dataset 2, solid lines with dataset 3. Reprinted with permission of The American Ceramic Society. **d** Calculated maximum contents of Al in β–SiAlON at various temperatures, dotted lines are calculated with dataset 2, solid lines with dataset 3. Reprinted with permission of The American Ceramic Society. **e** Isothermal section of Si_3N_4–AlN–Al_2O_3–SiO_2 system at 1600 °C, calculated by dataset 2. Reprinted from Ref. [35], Copyright 2007, with permission from Elsevier. **f** Isothermal section of Si_3N_4–AlN–Al_2O_3–SiO_2 system at 1800 °C, calculated by dataset 2. Reprinted from Ref. [35], Copyright 2007, with permission from Elsevier

Fig. 1.21 Isothermal section of Si_3N_4–Li_2O–Al_2O_3 pseudoternary system at 1550 °C. Reprinted with permission of The American Ceramic Society

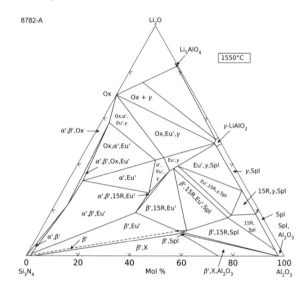

1.2.4 Si_3N_4–SiO_2–AlN–Al_2O_3–Be_3N_2–BeO

In the $(AlN)_4$–$(Al_2O_3)_2$–$(SiO_2)_3$–Si_3N_4–$(Be_3N_2)_2$–$(BeO)_6$ Jänecke System [38], Fig. 1.22a shows the schematic quaternary orientation diagram of Si_3N_4–(AlN: Al_2O_3)–$BeAl_2O_4$–Be_2SiO_4 system; Fig. 1.22b shows the isothermal section of Si_3N_4–(AlN:Al_2O_3)–$BeAl_2O_4$–Be_2SiO_4 system at 1760 °C, in which an area of β(Be)–SiAlON exists.

β(Be)–SiAlON is a substituted solid solution of $Si_{6-x-y}Be_xAl_yO_{2x+y}N_{8-2x-y}$ with M/X = 3/4. Except Al–O substitutes for Si–N forming SiAlON, many small metal elements are also able to enter into Si_3N_4 lattice to form β–$(M,Al,Si)_6(N,O)_8$ solid solution, i.e., (M)–SiAlON, M = Li, Be, Mg, Ga, and Ge.

Fig. 1.22   a Prism diagram of
$(AlN)_4$–$(Al_2O_3)_2$–$(SiO_2)_3$–
Si_3N_4–$(Be_3N_2)_2$–$(BeO)_6$
Jänecke system. Reprinted
with permission of The
American Ceramic Society.
b Isothermal section of
Si_3N_4–$(AlN{:}Al_2O_3)$–
$BeAl_2O_4$–Be_2SiO_4 quinary
system at 1760 °C. Reprinted
with permission of The
American Ceramic Society

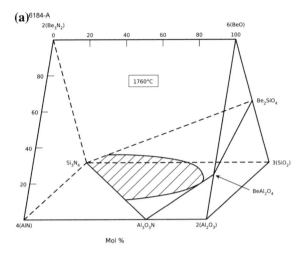

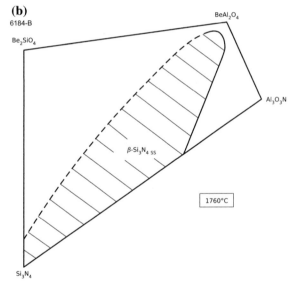

1.2.5 Si_3N_4–Al_2O_3–MgO

Figure 1.23 shows the phase diagram of MgO–Si_3N_4–Al_2O_3 system at 1800 °C, in which α = α-silicon nitride; β = β-Si_3N_4; β' = magnesium SiAlON; Spl = a defect structure, $Si_{11.5}N_{15}O_{0.5}$ ($MgAl_2O_4$); Per = periclase (MgO); Fo = forsterite (Mg_2SiO_4); 15R = phase of the 15R polytype of the AlN structure; 12H = SiAlON phase of the 12H polytype of the AlN structure; X = "nitrogen mullite" $\sim Al_6Si_6N_8O_9$; and N = "nitrogen spinel". The β' phase is said to have the

Fig. 1.23 a Phase diagram of
Si₃N₄–Al₂O₃–MgO
pseudoternary system.
Reprinted with permission of
The American Ceramic
Society. b Phase diagram of
Si₃N₄–Al₂O₃–MgO
pseudoternary system.
Reprinted with permission of
The American Ceramic
Society

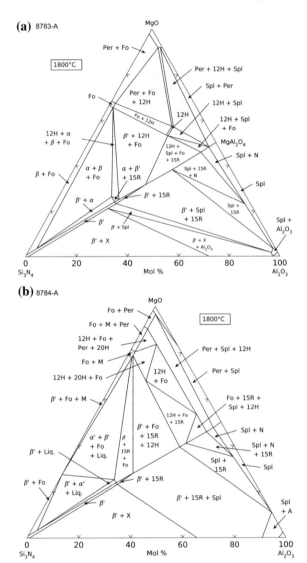

same structural unit as β–Si₃N₄ and may form β(Mg)–SiAlON with a formula of $Mg_{x/4}Si_{6-3x/4}Al_{x/2}O_xN_{8-x}$.

Figure 1.23a is from [36]; Fig. 1.23b shows the compatibility diagram of Si₃N₄–Al₂O₃–MgO pseudoternary system [37], in which β(Mg)–SiAlON forms on the Si₃N₄–MgAl₂O₄ (M/X = 3/4) line with the formula of $Mg_{x/4}Si_{6-3x/4}Al_{x/2}O_xN_{8-x}$.

1.2.6 Si_3N_4–SiO_2–AlN–Al_2O_3–MgO

Figure 1.24a shows the phase diagram of β– and β(Mg)–SiAlON relationships on the plane of M/X = (Mg,Al,Si)/(O,N) = 3/4 in the Si–Al–Mg–O–N system [37]; Fig. 1.24b shows the phase relations on the M/X = 3/4 plane determined from samples hot pressed at 1800 °C, in which liquid regions are surrounded by point line. The dotted line is, respectively, at 1500, 1600, and 1700 °C [12]; Fig. 1.24c shows the phase relations on the M/X = 3/4 plane determined from the samples hot pressed at 1750 °C [20].

On the plane of $3/2(Mg_2SiO_4)$–Si_3N_4–$4/3(Al_3O_3N)$–$3/2(MgAl_2O_4)$ along the ratio of M/X = (Mg,Al,Si)/(O,N) = 3/4, two solid solutions of β–SiAlON (Si_3N_4–AlN:Al_2O_3 ss) and β(Mg)–SiAlON (Si_3N_4–$MgAl_2O_4$ spinel ss) were formed. 15R and 21R polytypoids were also generated.

1.2.7 Ln_2O_3–Si_3N_4–AlN–Al_2O_3 (Ln = Nd, Sm)

Figure 1.25 shows the prism diagram of Ln–Si–Al–O–N (Ln = Sm, Nd) system at 1700 °C [39]. Subsolidus phase relationships have been determined, in the systems of Ln–Si–Al–O–N where Ln = Nd, Sm. Forty-four compatibility tetrahedra were established in the region of $Ln_2O_3 \sim Si_3N_4$–AlN–Al_2O_3. Within this region, $LnAlO_3$ and M′-phase ($Ln_2Si_{3-x}Al_xO_{3+x}N_{4-x}$) are the only two important compounds which have tie lines joined to β–SiAlON and AlN polytypoid phases. α–SiAlON coexists with the M′-phase. Five phase diagrams of these two Ln–Si–Al–O–N (Ln = Sm, Nd) systems are collected here.

1.2.8 Si_3N_4–AlN–Al_2O_3–Y_2O_3

Figure 1.26 shows a tentative subsolidus diagram of Y_2O_3–Al_3O_3N–Si_3N_4 system [40], in which YAM′ = N-containing $Y_4Al_2O_9$, YAG′ = N-containing $Y_3Al_5O_{12}$, J′ = $\sim Y_4Si_2O_7N_2$, and $β_{60}$ = limiting SiAlON ss with 60 eq.% Al.

Starting powders were Al_2O_3 (99.99% pure), AlN (combined O 1.4%), Si_3N_4 (combined O 1.2%), and Y_2O_3 (99.9% pure). When weighing these materials to prepare samples, the oxygen impurity was taken into account. Mixtures for 36 compositions were ground in an agate mortar in water-free alcohol for 2 h, then dried and isostatically pressed at ~ 0.4 GPa. Pellets were contained in BN-lined graphite crucibles filled with AlN powder and fired under "mildly flowing" N_2 for 1/2–3 h at 1550–1750 °C. In order to confirm equilibrium, most compositions were fired or hot pressed under two different temperature/holding time conditions, for example, fired at 1550 °C/3 h or hot pressed at 1750 °C/0.5 h. Samples with weight losses greater than 3 wt% were not used.

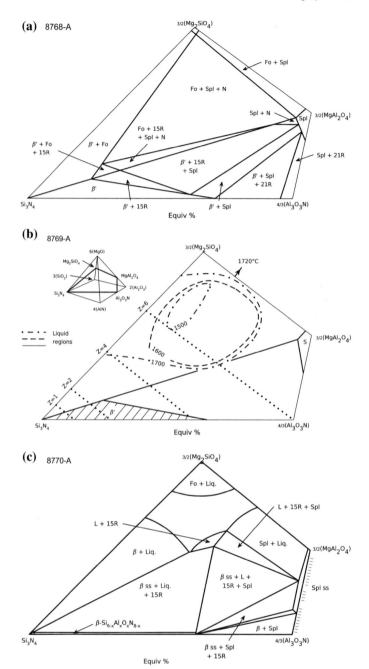

Fig. 1.24 a The phase diagram of β– and β(Mg)–SiAlON relationships. Reprinted with permission of The American Ceramic Society. **b** Phase relations on the M/X = 3/4 plane determined from the samples hot pressed at 1800 °C. Reprinted with permission of The American Ceramic Society. **c** Phase relations on the M/X = 3/4 plane determined from the samples hot pressed at 1750 °C. Reprinted with permission of The American Ceramic Society

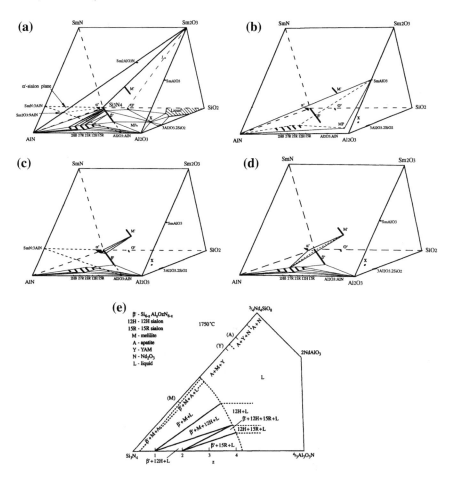

Fig. 1.25 **a** Si$_3$N$_4$–Sm(Nd)$_2$O$_3$–Al$_2$O$_3$–AlN region in prism of Ln–Si–Al–O–N (Ln = Sm, Nd) at 1700 °C. Reprinted from Ref. [39], Copyright 1995, with permission from Elsevier. **b** Polytypes in Sm(Nd)AlO$_3$ and Si-rich area, five compatible tetrahedrons formed from AlN and Sm(Nd) Al12O18N (MP compounds). Reprinted from Ref. [39], Copyright 1995, with permission from Elsevier. **c** α′–β′–M′ compatible tetrahedrons formed from M′(melilite ss), β–SiAlON (β$_0$–β$_{10}$) and α–SiAlON. Reprinted from Ref. [39], Copyright 1995, with permission from Elsevier. **d** Compatible tetrahedron of α′–β′–21R–M′. Reprinted from Ref. [39], Copyright 1995, with permission from Elsevier. **e** Phase relation of β′–NdAl0$_3$ plane at 1750 °C. Reprinted from Ref. [39], Copyright 1995, with permission from Elsevier

The authors use the symbol J′ for the compound Y$_4$Si$_2$N$_2$O$_7$ as it is not a phase of the binary indicated but was always observed in this work, probably due to some oxidation during sintering. YAM′ and J′ are isostructural (monoclinic) phases and form continuous solid solution in between.

Fig. 1.26 Tentative
subsolidus diagram of Y$_2$O$_3$–
Al$_3$O$_3$N–Si$_3$N$_4$ system.
Reprinted with permission of
The American Ceramic
Society

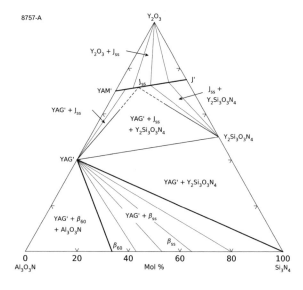

1.2.9 Si$_3$N$_4$–SiO$_2$–AlN–Al$_2$O$_3$–Y$_2$O$_3$

Figure 1.27 shows a Jänecke prism of Si$_3$N$_4$–3SiO2–4AlN–2Al$_2$O$_3$–4YN–2Y$_2$O$_3$
[41], in which β' ss = Al$_x$Si$_{3-x}$O$_x$N$_{4-x}$, $0 \leq x \leq 2$ and "β_{10}" =
Al$_{0.3871}$Si$_{2.613}$O$_{0.3871}$N$_{3.613}$. Over 50 compositions were fired at 1550 °C/1 h. Those
containing glass were further annealed at 1150–1250 °C for 24 h to determine true
subsolidus equilibria. The phase compositions of samples were analyzed using XRD.
Figure 1.27a shows the positions of solid phases in the prism diagram of Si–Al–Y–O–
N system. The important compatibility tetrahedra are shown in Fig. 1.27b, c. Eighteen
other compatibility tetrahedra exist and are listed here. As indicated in Fig. 1.27a,
there are X = Al$_{18}$Si$_{12}$O$_{39}$N$_8$ and H = Y$_5$Si$_3$O$_{12}$N; β' ss and Si$_2$ON$_2$ ss are solid
solutions.

1. Al$_2$SiO$_2$N$_2$–Y$_2$SiO$_5$–X–β_{10},
2. Al$_2$SiO$_2$N$_2$–Y$_2$SiO$_5$–X–Y$_3$Al$_5$O$_{12}$,
3. Al$_2$SiO$_2$N$_2$–Al$_2$O$_3$–X–Y$_3$Al$_5$O$_{12}$,
4. X–SiO$_2$–Y$_2$Si$_2$O$_7$–Si$_2$ON$_2$ ss,
5. X–Y$_2$Si$_2$O$_7$–Si$_2$ON$_2$ ss–β_{10},
6. Y$_2$Si$_2$O$_7$–(Si$_2$ON$_2$ ss–Si$_2$ON$_2$)–SiO$_2$,
7. {Si$_3$N$_4$–β_{10}}–{Si$_2$ON$_2$ ss–Si$_2$ON$_2$}–Y$_2$Si$_2$O$_7$,
8. Si$_3$N$_4$–Y$_3$Al$_5$O$_{12}$–Y$_2$Si$_2$O$_7$–H,
9. Si$_3$N$_4$–H–Y$_3$Al$_5$O$_{12}$–Y$_2$Si$_2$O$_4$N$_2$,
10. Si$_3$N$_4$–Y$_2$Si$_2$O$_4$N$_2$–Y$_2$Si$_3$O$_3$N$_4$–Y$_3$Al$_5$O$_{12}$,
11. Y$_2$Si$_2$O$_4$N$_2$–Y$_2$Si$_3$O$_3$N$_4$–Y$_4$Al$_2$O$_9$–Y$_3$Al$_5$O$_{12}$,
12. Y$_2$Si$_2$O$_4$N$_2$–Y$_3$Al$_5$O$_{12}$–Y$_4$Al$_2$O$_9$–H,

(a) 8776-A

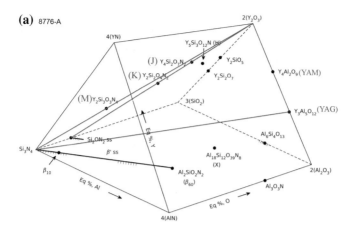

(b) 8776-B

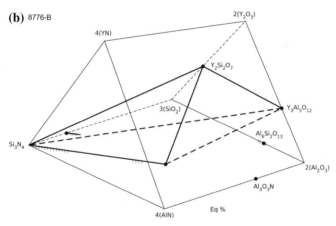

(c) 8776-C

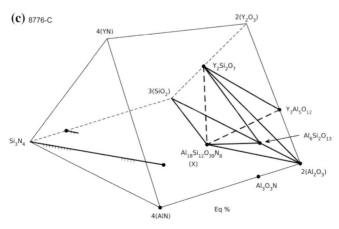

◄**Fig. 1.27** **a** Jänecke prism of Si–Al–Y–O–N system, showing compound locations. Reprinted with permission of The American Ceramic Society. **b** Compatibility tetrahedra of β–SiAlON–YAG–$Y_2Si_2O_7$. Reprinted with permission of The American Ceramic Society. **c** Compatibility tetrahedra of YAG–X–$Y_2Si_2O_7$–Al_2O_3. Reprinted with permission of The American Ceramic Society

13. H–$Y_3Al_5O_{12}$–$Y_2Si_2O_7$–Y_2SiO_5,
14. H–$Y_3Al_5O_{12}$–Y_2SiO_5–$Y_4Al_2O_9$,
15. $Y_2Si_3O_3N_4$–$Y_2Si_2O_4N_2$–$Y_4Si_2O_7N_2$–$Y_4Al_2O_9$,
16. $Y_2Si_2O_4N_2$–H–$Y_4Si_2O_7N_2$–$Y_4Al_2O_9$,
17. $Y_4Si_2O_7N_2$–H–$Y_4Al_2O_9$–Y_2O_3, and
18. H–$Y_4Al_2O_9$–Y_2SiO_5–Y_2O_3.

1.2.10 Si_3N_4–$Y_{12}Si_5O_{28}$–$Y_3Al_5O_{12}$–Al_3O_3N

Figure 1.28 shows the subsolidus diagram of Si_3N_4–$Y_{12}Si_5O_{28}$–$Y_3Al_5O_{12}$–Al_3O_3N plane separated from the Si_3N_4–Y_2O_3–SiO_2–Al_2O_3–AlN–YN system [42], at (a) 1700 °C, (b) 1350 °C, and (c) 1050 °C. Figure 1.28d shows the behavior in the plane Si_3N_4–Al_3NO_3–YAl_3N_4 at 1750 °C. In this figure, $\alpha' = \alpha'$–Si_3N_4 ss; $\beta' = \beta'$–SiAlON ss; YAG = $Y_3Al_5O_{12}$; N–Ap = $Y_{10}(SiO_4)_6N_2$; N–Wo = $YSiO_2N$; J = $Y_4Al_2O_9$–$Y_4Si_2O_7N_2$ ss; P = polytypoid phase; YS = Y_2SiO_5; YS_2 = $Y_2Si_2O_7$; B ≈ Y_2SiAlO_5N ss; 15R, 12H, 21R, and 27R are AlON polytypoids; and M = $Y_2Si_3O_3N_4$.

The paper [42] presented a review of phase relationships in ceramics based on reactions between Si_3N_4, YN, AlN, SiO_2, Y_2O_3, or Al_2O_3 between 1050 and 1750 °C. X-ray powder patterns were presented for Y_3O_3N, YSi_3N_5, $Y_2Si_3N_6$, $Y_6Si_3N_{10}$, and $Y_4Al_2O_9$ (YAM-type), and lattice parameter values for solid solutions between $YSiO_2N$ and $YAlO_3$, and $Y_4Si_2O_7N_2$ (J-type) and $Y_4Al_2O_9$ (YAM-type) were plotted. See PED 6181C for an earlier version of Fig. 1.28a. The author [42] compared Fig. 1.28a with PED 6181E and pointed out the need for careful specimen preparation, and the difficulty in devitrifying glassy phases and in ensuring equilibrium. All phases in Fig. 1.28a are stable at temperatures up to 1600 °C, except for the N–α-wollastonite which decomposes above ∼1400 °C to $Y_{10}(SiO_4)_6N_2$ and $Y_2Si_3O_3N_4$.

β'–SiAlON is in equilibrium only with $Y_3Al_5O_{12}$ in this system at higher temperatures. At lower temperatures, the glassy phase in the sample devitrifies to phase "B", nominally Y_2SiAlO_5N, a wollastonite-type phase. Figure 1.28a–c shows the phases formed at various temperatures and illustrates the fact that subsolidus phase relationships may change depending on devitrification temperature and the difficulty in preparing β'–YAG composites free of other crystalline phases. Figure 1.28d shows that in contrast to β'–SiAlON, the yttrium α'–SiAlON is in equilibrium only with the line of compositions facing the β', and the range of liquid

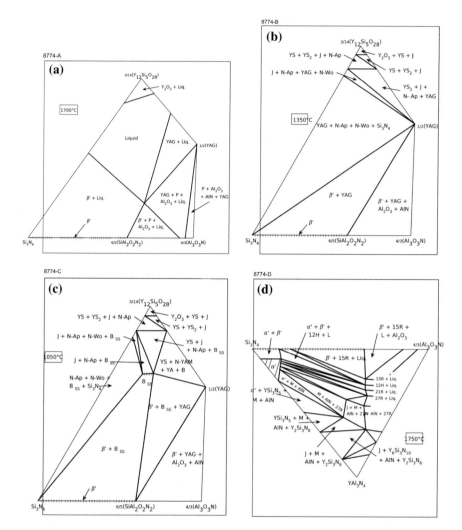

Fig. 1.28 a Isothermal section of Si_3N_4–$Y_{12}Si_5O_{28}$–$Y_3Al_5O_{12}$–Al_3O_3N system at 1700 °C. Reprinted with permission of The American Ceramic Society. **b** Isothermal section of Si_3N_4–$Y_{12}Si_5O_{28}$–$Y_3Al_5O_{12}$–Al_3O_3N system at 1350 °C. Reprinted with permission of The American Ceramic Society. **c** Isothermal section of Si_3N_4–$Y_{12}Si_5O_{28}$–$Y_3Al_5O_{12}$–Al_3O_3N system at 1050 °C. Reprinted with permission of The American Ceramic Society. **d** Behavior diagram of Si_3N_4–$4/3Al_3NO_3$–YAl_3N_4 plane at 1750 °C. Reprinted with permission of The American Ceramic Society

compositions is very restricted. Polytypoid phases occur readily so that it is impossible to prepare α′–SiAlONs containing a single crystalline phase by devitrification. The whole Jänecke prism of this system is shown in PED 8773.

1.2.11   Si$_3$N$_4$–4(YN)–4(AlN)–2(Al$_2$O$_3$)–3(SiO$_2$)–2(Y$_2$O$_3$)

Figure 1.29 shows the phase diagrams of 4(YN)–4(AlN)–2(Al$_2$O$_3$)–3(SiO$_2$)–2 (Y$_2$O$_3$)–Si$_3$N$_4$ system [43]:

(a) Phase relations of SiAlON plane at 1700 °C in the Si–Al–Y–O–N Jänecke prism;

(b) Compatibility relations of β–SiAlON–α′–SiAlON–YAG–12H;

(c) Compatibility relations of β–SiAlON–α′–SiAlON–AlN polytype; and

(d) Phase relations in nitrogen-rich area of Si$_3$N$_4$–AlN–YN–Y$_2$O$_3$ system.

In Fig. 1.29, $β' = Si_{3-x}Al_xO_xN_{4-x}$ ss, $β_{10}$ and $β_{60}$ are β′ ss with $3x/[4 (3 − x) + 3x]$ (Al^{3+} equivalence), respectively, $β = 0.10$ and 0.60 (i.e., $x = 0.3871$, and 2.0); $X =$ "N-mullite" Al$_6$Si$_6$O$_9$N$_8$; 2Hδ, 27R, 21R, 12H, and 15R are SiAlON polytypes, as described in [43]; M = melilite, Y$_2$Si$_3$O$_3$N$_4$; J = Cuspidine (group of Wohlerite), Y$_4$Si$_2$O$_7$N$_2$; Jss = "Y$_4$Si$_2$O$_7$N$_2$–Y$_4$Al$_2$O$_9$" ss; O′ = Si$_2$ON$_2$-type SiAlON; H = apatite Y$_{10}$(SiO$_4$)$_6$N$_2$; K = wollastonite, YSiO$_2$N; and YAM = Y-Wohlerite, Y$_4$Al$_2$O$_9$.

Fifty-one compositions were studied. Thirty-three of them lie in the area without YN, and eighteen lie in the Si$_3$N$_4$–AlN–YN–Y$_2$O$_3$ area. Samples were fired in N$_2$ atmosphere at 1550–1900 °C (for compositions with YN, >1700 °C) for 1–2 h in a

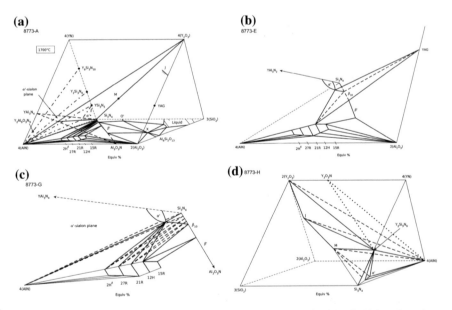

Fig. 1.29 **a** Phase relations of SiAlON plane at 1700 °C in the Si–Al–Y–O–N Jänecke prism. Reprinted with permission of The American Ceramic Society. **b** Compatibility relations of β–SiAlON–α′–SiAlON–YAG–12H. Reprinted with permission of The American Ceramic Society. **c** Compatibility relations of β–SiAlON–α′–SiAlON–AlN polytype. Reprinted with permission of The American Ceramic Society. **d** Phase relations in nitrogen-rich area of Si$_3$N$_4$–AlN–YN–Y$_2$O$_3$ system. Reprinted with permission of The American Ceramic Society

graphite furnace. YN powder was prepared by $Y_2O_3 + 3C + N_2(g) = 2YN + 3CO$ (g) reaction. In [42] (Fig. 1.28), three nitrides $Y_6Si_3N_{10}$, $Y_2Si_3N_6$, and YSi_3N_5 (see PED 8773 (A)) were reported, but only $Y_2Si_3N_6$ was notarized in the N-rich area (see PED8773(H)) [43]. β'–SiAlON was coexistent with all AlN polytypoids [44, 45]. α'–SiAlON plane was shown in PED8774(D) by a dot on the Si_3N_4–Al_3O_3N–YAl_3N_4 plane, although it has a large area (see PED 8775(D)). This work established 68 compatibility tetrahedra, in which 39 tetrahedra were in the Si_3N_4–SiO_2–Al_2O_3–Y_2O_3 region and 7 in the Si_3N_4–AlN–YN–Y_2O_3 region. Other 22 compatibility tetrahedrals were reported in PED8776, also seen in PED 8774.

1.2.12 Si_3N_4–Al_4N_4–Y_4O_6

Starting powders were Si_3N_4, AlN, and Y_2O_3. The samples were fired at 1750 °C for 2 h in 1 MPa N_2 [6]. In Fig. 1.30, the phase diagram of Si_3N_4–Al_4N_4–Y_4O_6 system in the Y–Si–Al–O–N prism was determined. Liquidus surfaces of Si_3N_4–Al_4N_4–Y_4O_6 system were of special significance for preparation of nitride ceramics using AlN–Y_2O_3 additives to promote liquid phase densification. The prism could be separated into two parts: nitride-rich region (upside) and oxide-rich region (downside) with Al_2O_3 and SiO_2. As well known, the ternary Y_2O_3–Al_2O_3–SiO_2 system has a low eutectic temperature of 1360 °C [46] and the SiAlON Si_3N_4–AlN–Al_2O_3–SiO_2 plane has T_{eu} 1480 °C [31]. Therefore, the down side is a lower eutectic temperature region. The upside part, i.e., the nitride-rich area of Si_3N_4–AlN–Y_2O_3 without Al_2O_3 and SiO_2, has a eutectic point T_{eu} = 1650 °C at the composition of $15Si_3N_4 + 25AlN + 60Y_2O_3$ (mol%), which was favorable for liquid phase sintering.

Fig. 1.30 Phase diagram of Si_3N_4–Al_4N_4–Y_4O_6 system at 1750 °C/1 MPa N_2. Reprinted from Ref. [6], Copyright 1996, with permission from John Wiley and Sons

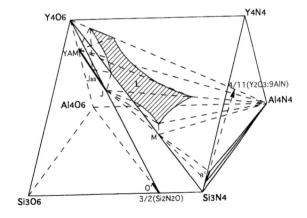

1.3 α–SiAlON System

The lattice of α–Si$_3$N$_4$ contains four molecules and seals two interstices inside, which permit foreign metal ions to fill in. When Al–O and Al–N substitute for Si–N in Si$_3$N$_4$, followed by other large metal ions, such as Na, Mg, Ca, Y and all rare earths except La and Ce, to enter into the interstices, the lattice structure will not be changed to form β–SiAlON at high temperatures. Instead, it will keep the same structure as α–Si$_3$N$_4$ to form a solid solution, named as α(M)–SiAlON, which is stable at high temperatures. Its molecular formula is $M_x Si_{12-(m+n)} Al_{(m+n)} O_n N_{16-n}$, as m(Al–N) and n(Al–O) displace for (m + n)Si–N, while the metal ions (M) occupy the two interstices at $x \leq 2$.

1.3.1 Si$_3$N$_4$–AlN–R$_2$O$_3$ System

When metal ion M = Rare earth, such as α(Y)–SiAlON [46, 47], the composition will be located in the Si$_3$N$_4$–Al$_3$O$_3$N–YAl$_3$N$_4$ triangle (see Fig. 1.31a, b, PED8772A and PED 8775 A and B) [46, 47]. The two-dimensional composition is $Y_{m/3} Si_{12(m+n)} Al_{(m+n)} O_n N_{16n}$, $n = 0$, $1.3 < m < 2.4$ to $n = 1.7$, $m = 1$. The end-member YAl$_3$N$_4$ contains YN, a very easy to be oxidized yttrium nitride, a situation that causes difficult processing in manufacture of ceramics.

The section between the two triangles of Si$_3$N$_4$–AlN–Y$_2$O$_3$ and Si$_3$N$_4$–Al$_3$O$_3$N–YAl$_3$N$_4$ on the α–SiAlON plane is a one-dimensional line, i.e., Si$_3$N$_4$–Y$_2$O$_3$:9AlN line (see Fig. 1.31c, and PED8754). The solid solubility of Y in α(Y)–SiAlON is 0.33–0.67 [48]. Within Y$_2$O$_3$:9AlN, except for the interstitial Y, 9Al/ (9N + 3O) = 9/12 = 3/4, the same ratio as Si/N = 3/4 in Si$_3$N$_4$. This is the prerequisite of substitution for each other between Si$_3$N$_4$ and 1/3(Y$_2$O$_3$:9AlN). For other rare earth, the lowest solid solubility x of R in all α(R)–SiAlONs is ~0.33 ions in every lattice. The maximum solid solubility x of R in α(R)–SiAlONs increases with decreasing ionic radius in the range $x = 0.6$ (for Nd)–1.0(for Yb) (see Fig. 1.31c–k) [49]. Such a one-dimensional composition line of Si$_3$N$_4$–R$_2$O$_3$:9AlN will benefit compositional design and easier processing of α(R)–SiAlON ceramics by avoiding usage of RN.

Figure 1.31 shows the 11 selected phase diagrams of α(R)–SiAlON systems.

1.3.2 Si$_3$N$_4$–AlN–M$_2$O, –M'O System

When metal ion M = Na^{1+}, α(Na)–SiAlON could be formed but unstable. As a transition phase, it is difficult to get single-phase α(Na)–SiAlON [50]. Its decomposition products were β(Na)–SiAlON, AlN, and 15R polytypoid. In process of

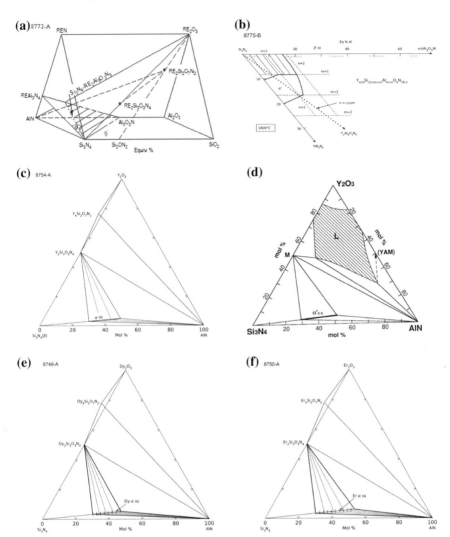

Fig. 1.31 Phase diagrams of α(R)–SiAlON systems: **a** Tri-prism of system RN–AlN–Si₃N₄–SiO₂–Al₂O₃–R₂O₃. Reprinted with permission of The American Ceramic Society. **b** Amplified SiAlON plane of 4(YN)–4(AlN)–2(Al₂O₃)–3(SiO₂)–2(Y₂O₃)–Si₃N₄ system. Reprinted with permission of The American Ceramic Society. **c** Subsolidus phase diagram of AlN–Si₃N₄–Y₂O₃ system. Reprinted with permission of The American Ceramic Society. **d** Phase diagram of Si₃N₄–Al₄N₄–Y₄O₆ system at 1750 °C/1 MPa N₂. Reprinted from Ref. [6], Copyright 1996, with permission from John Wiley and Sons. **e–j** Phase diagrams of AlN–Si₃N₄–R₂O₃ systems, R = Dy, Er, Gd, Nd, Sm, and Yb. Reprinted with permission of The American Ceramic Society. **k** Solid solubility of rare earth ions (X) in α–Rₓ(Si,Al)₁₂(O,N)₁₆ with different ionic radiuses. Reprinted from Ref. [46], Copyright 1986, with permission from John Wiley and Sons

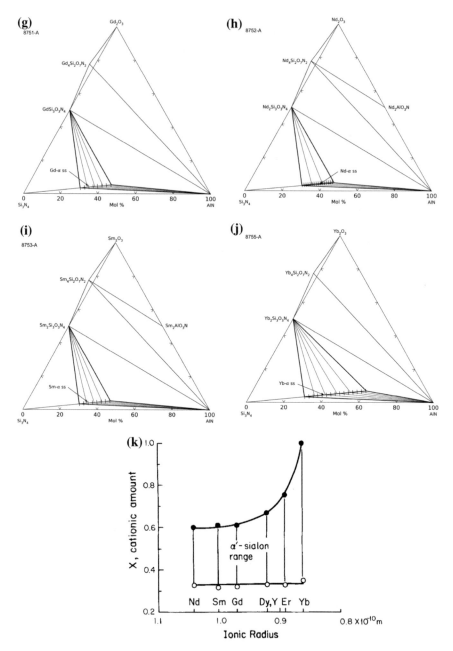

Fig. 1.31 (continued)

forming α(Na)–SiAlON, a new transition phase, O′(Na)–SiAlON, was also observed. Figure 1.32a shows the tentative α(Na)–SiAlON plane.

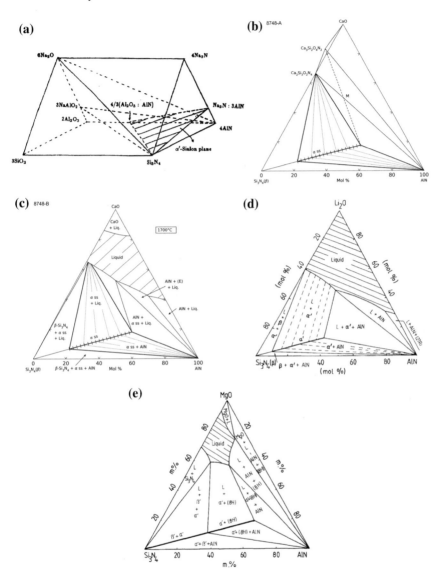

Fig. 1.32 a α(Na)–SiAlON plane in the Si–Al–Na–O–N system. Reprinted from Ref. [50], Copyright 1994, with permission from Journal of The Inorganic Materials (Chinese). **b** Subsolidus phase diagram of the AlN–Si₃N₄–CaO system. Reprinted with permission of The American Ceramic Society. **c** Phase diagram of the AlN–Si₃N₄–CaO system at 1700 °C. Reprinted with permission of The American Ceramic Society. **d** Phase diagram of the AlN–Si₃N₄–Li₂O system. Reprinted from Ref. [52], Copyright 1990, with permission from Springer Nature. **e** Phase diagram of the AlN–Si₃N₄–MgO system. Reprinted from Ref. [53], Copyright 1990, with permission from Springer Nature

For M' = Ca^{2+}, α single-phase α(Ca)–SiAlON is very easy to form by solid-state reaction of compositions on the Si$_3$N$_4$–CaO:3AlN one-dimensional line (see Fig. 1.32b, c). The solid solubility of α(Ca)–SiAlON is high, n = 0.3–1.4 [51] in Ca$_n$Si$_{12-3n}$Al$_{3n}$O$_n$N$_{16-n}$. Except for α(Ca)–SiAlON in the AlN–Si$_3$N$_4$–CaO system, there exist two other compounds, i.e., (1) Ca$_3$Si$_2$O$_4$N$_2$ with cubic structure (a = 1.507 nm) and the melting point of ∼1580 °C, and (2) Ca$_2$Si$_3$O$_2$N$_4$ with unknown structure and the melting point of ∼1680 °C. A nitride CaAlSiN$_3$ (E phase) with orthorhombic structure was also absorbed but did not belong to this ternary system. It may be formed by decomposition of "M" phase [51] (see Fig. 1.32b).

For M = Li^{1+}, the formation of single-phase α(Li)–SiAlON with a chemical formula of Li$_{2n}$Si$_{12-3n}$Al$_{3n}$O$_n$N$_{16-n}$ was observed on the one-dimensional composition line of Si$_3$N$_4$–Li$_2$O:3AlN. It was evidenced to have a high solid solubility of 2n = 0.25–0.5 (see Fig. 1.32d). The lattice parameters of hexagonal α(Li)–SiAlON increase with increasing 2n, with the maximum a = 0.7836 nm and c = 0.5687 nm [52].

When M' = Mg^{2+}, α(Mg)–SiAlON exists with compositions along the one-dimensional line of Si$_3$N$_4$–MgO:3AlN. But it is difficult for single-phase α(Mg)–SiAlON to form, and α(Mg)–SiAlON often coexist with β–SiAlON and Mg polytypoid 8H (see Fig. 1.32e) [53].

In Fig. 1.32, five phase diagrams of the Si$_3$N$_4$–AlN–M$_2$O, –M'O systems were selected.

1.4 O′–SiAlONs

O′–SiAlON is a Si$_2$ON$_2$–Al$_2$O$_3$ solid solution formed by partially substituting Al–O in Al$_2$O$_3$ for Si–N in Si$_2$ON$_2$. The Si$_2$ON$_2$ ss has a chemical formula of Si$_{2-x}$Al$_x$O$_{1+x}$N$_{2-x}$ (0 ≤ x ≤ 0.4). It is named as O′–SiAlON for its excellent oxidation resistance.

1.4.1 Si$_2$ON$_2$–Al$_2$O$_3$–M$_x$O$_y$

In Si$_2$N$_2$O–Al$_2$O$_3$–M$_x$O$_y$ (M = Ca, Mg, La, Y) systems, O′(M)–SiAlON coexist with either β–Si$_3$N$_4$ or β–SiAlON (Fig. 1.33) [54–57]. O′(M)–SiAlON also coexist with CaSiO$_3$ in Ca-system, MgAl$_2$Si$_4$O$_6$N$_4$ (N phase) in Mg-system, La$_5$Si$_3$O$_{12}$N (La-apatite) in La-system, Y$_5$Si$_3$O$_{12}$N (Y-apatite), and Y$_3$Al$_5$O$_{12}$ (YAG) in Y-system. In addition, Ca$_3$Si$_2$O$_4$N$_2$–Ca$_3$Al$_2$O$_6$ in Ca-system and Y$_4$Si$_2$O$_7$N$_2$ (J-phase)-Y$_4$Al$_2$O$_9$ (YAM) in Y-system form respective continuous solid solution.

Figure 1.33 shows the phase diagrams of Si$_2$ON$_2$–Al$_2$O$_3$–M$_x$O$_y$ (M = Ca, La, Y) systems:

(a) Phase diagram of Y$_2$O$_3$–Si$_2$ON$_2$–Al$_2$O$_3$ system at 1400 °C [54],

(b) Subsolidus compatibility of Si$_2$ON$_2$–CaO–Al$_2$O$_3$ pseudoternary system [55],

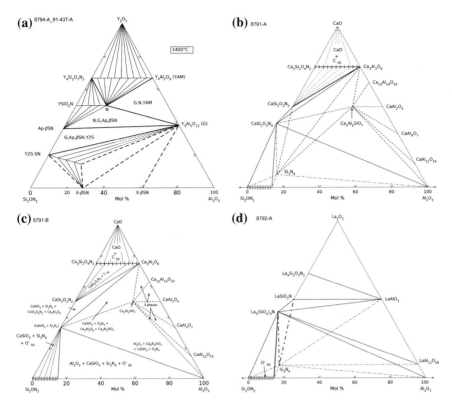

Fig. 1.33 a Phase diagram of Y_2O_3–Si_2ON_2–Al_2O_3 system at 1400 °C. Reprinted with permission of The American Ceramic Society. **b** Subsolidus compatibility of Si_2ON_2–CaO–Al_2O_3 pseudoternary system. Reprinted with permission of The American Ceramic Society. **c** Projection plane of (**b**) of Si_2ON_2–CaO–Al_2O_3 pseudoternary system. Reprinted with permission of The American Ceramic Society. **d** Subsolidus compatibility of Si_2ON_2–La_2O_3–Al_2O_3 pseudoternary system. Reprinted with permission of The American Ceramic Society. **e** Projection plane of (**d**) of Si_2ON_2–La_2O_3–Al_2O_3 pseudoternary system. Reprinted with permission of The American Ceramic Society. **f** Subsolidus compatibility of Si_2ON_2–Y_2O_3–Al_2O_3 pseudoternary system. Reprinted with permission of The American Ceramic Society. **g** Projection plane of (**f**) of Si_2ON_2–Y_2O_3–Al_2O_3 pseudoternary system. Reprinted with permission of The American Ceramic Society. **h** Isothermal section of Si_2ON_2–Y_2O_3–Al_2O_3 pseudoternary system at 1550 °C. Reprinted with permission of The American Ceramic Society

(c) Projection plane of Fig. 1.33b of Si_2ON_2–CaO–Al_2O_3 pseudoternary system [56],

(d) Subsolidus compatibility of Si_2ON_2–La_2O_3–Al_2O_3 pseudoternary system [55],

(e) Projection plane of Fig. 1.33d of Si_2ON_2–La_2O_3–Al_2O_3 pseudoternary system [56],

(f) Subsolidus compatibility of Si_2ON_2–Y_2O_3–Al_2O_3 pseudoternary system [57], and

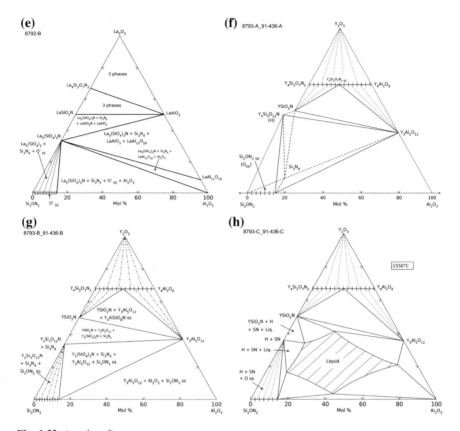

Fig. 1.33 (continued)

(g) Projection plane of Fig. 1.33f of Si₂ON₂–Y₂O₃–Al₂O₃ pseudoternary system [57], and

(h) Isothermal section of Si₂ON₂–Y₂O₃–Al₂O₃ pseudoternary system at 1550 °C [57].

1.5   M′(R)–SiAlONs System

M′(R)–SiAlON is a R₂O₃·Si₃N₄ (R–melilite) solid solution which is formed by partial substitution of Al–O in Al₂O₃ for Si–N in R-melilite.

1.5.1   Si₃N₄–R₂O₃–Al₂O₃ (M′(R)–SiAlONs)

In this ternary Si₃N₄–R₂O₃–Al₂O₃ system, all rare earth systems form R₂O₃·Si₃N₄ (tetragonal R-melilite) except for La. R-melilite, except for Yb- and Lu-melilite, can be partially substituted by Al_2O_3 to form melilite solid solutions of M′(R)–SiAlONs with a chemical formula $R_2Si_{3-x}Al_xO_{3+x}N_{4-x}$ at 1 (for La) > x > 0.6 (for Er) [58, 59]. The solid solubility of M′(R)–SiAlON decreases with shrinking radius of rare earth ion. The dependence of lattice parameters on x (i.e., Al concentration) and r (i.e., radius of rare earth ion in Å from Ahren) is determined as

a (Å) = 6.795 + 0.892r +0.045x and

c (Å) = 4.121 + 0.874r +0.033x.

La has the biggest ionic radius. La can enter melilite lattice only at the largest substitution of x = 1 to form M′(La) with a fixed composition of $La_2Si_2AlO_4N_3$. It is a tetragonal compound with lattice parameters a = 7.835, c = 5.120 Å, and can be written as $La_2O_3\cdot Si_2N_2O\cdot AlN$.

Figure 1.34 shows the equilibria relationships of M′(R) ss with neighboring phases [59]. M′(La) is denoted by a dot in Fig. 1.34a for that it is either an M′(La) ss with x = 1 or a compound. The M(R)–M′(R) ss tie line shrinks in length from Nd, Sm in Fig. 1.34b to Gd, Dy, Er, and Y in Fig. 1.34c, further to M(Yb) as a dot in Fig. 1.34d for the reason that M(Yb) has no solubility.

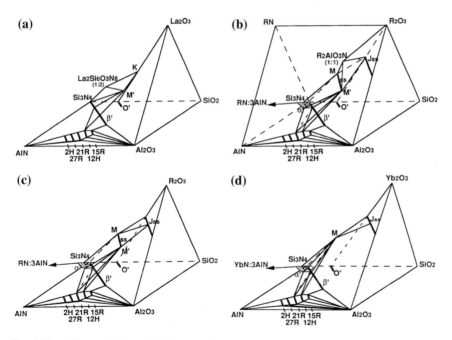

Fig. 1.34 a Phase relations of M′(La) with neighboring phases in La–Si–Al–O–N system. Reprinted from Ref. [59], Copyright 1996, with permission from John Wiley and Sons. **b** Phase relations of M′(R)(R = Nd, Sm) with neighboring phases in R–Si–Al–O–N system. Reprinted from Ref. [59], Copyright 1996, with permission from John Wiley and Sons. **c** Phase relations of M′(R) (R = Gd, Dy, Er, Y) with neighboring phases in R–Si–Al–O–N system. Reprinted from Ref. [59], Copyright 1996, with permission from John Wiley and Sons. **d** Phase relations of M′ (Yb) with neighboring phases in Yb–Si–Al–O–N system. Reprinted from Ref. [59], Copyright 1996, with permission from John Wiley and Sons

1.6   AlSiONs (SiAlON Polytypoids)

There exist a series of AlN polytypes in the AlN–Al_2O_3 binary system (see
PED9107). They will be Si-containing AlN polytypoids while their compositions
enter into the SiAlON plane in the Si–Al–O–N system. Different from the other
SiAlONs, the AlN polytypoid can be referred as AlSiON (i.e., Al–Si–O–N) polytypoids or named as SiAlON polytypoid with AlN structure.

1.6.1   SiAlONs Polytypoids

Similar to the SiC polytypes, the AlN polytypoids have two categories of structures,
i.e., Hexagonal (*n*H) and Rhombohedral (*n*R). They are marked as *n*H and *n*R by
the Ramsdell's symbol system, respectively, where the number *n* refers to the
identical atomic layers in one lattice with periodical stacking. For the *n*H polytypoid, a lattice unit contains two base blocks; each block is consisted of *n*/2 atomic
layers. For the *n*R polytypoid, a lattice unit contains three blocks; each block is
consisted of *n*/3 atomic layers. Table 1.1 lists the compositions of the SiAlON
polytypoids [44].

In the Si–Al–Y–O–N system, all AlN polytypoids could coexist with each of
α–SiAlON, β–SiAlON, and YAG. Figure 1.35 shows $Y_3Al_5O_{12}$ (YAG) and

Table 1.1 Compositions of SiAlON polytypoids [44]

Polytypoids	Substitute atoms (x)	Cations/anions ratio	Molecules
8H	5	4/5	$Si_{6-x}Al_{2+x}O_xN_{10-x}$
15R	4–5	5/6	$Si_{6-x}Al_{4+x}O_xN_{12-x}$
12H	4.5–6	6/7	$Si_{6-x}Al_{6+x}O_xN_{14-x}$
21R	4.5–6	7/8	$Si_{6-x}Al_{8+x}O_xN_{16-x}$
27R	4–6	9/10	$Si_{6-x}Al_{10+x}O_xN_{20-x}$
2H^9	(2?)–6	11/12	$Si_{6-x}Al_{16+x}O_xN_{24-x}$

Fig. 1.35 Compatibility of
AlN polytypoids with YAG.
Reprinted with permission of
The American Ceramic
Society

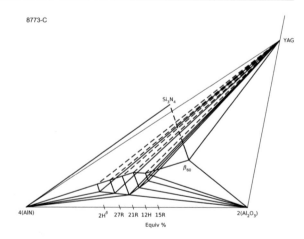

AlSiONs in equilibrium [43]. The AlN polytypoids with wurtzite structure (i.e., nH) are often used as the secondary phase in composite ceramics to play an enforcing role and promote toughness.

1.6.2 Si_3N_4–$(Be_3N_2)_2$–$(BeO)_6$–$(SiO_2)_3$

Figure 1.36 shows the isothermal section of Si_3N_4–$(Be_3N_2)_2$–$(BeO)_6$–$(SiO_2)_3$ system at 1780 °C [60]. Except for AlN polytypoids, when Be and Mg participate, BeSiON and MgSiON polytypoids are formed, respectively. At the Be_3N_2-rich corner of the Si_3N_4–SiO_2–BeO–Be_3N_2 system, BeSiON polytypoids are formed to have more 4H and 9R than that of AlN polytypoids. For MgSiON [37], there are 6H, 8H, 12H, 21R, 16H, and 27R polytypoids.

   Huseby et al. [60] prepared 52 samples with different compositions by hot pressing at 1765 °C–1880° C for 1–2 h under 28 MPa pressure. XRD was used to analyze phase compositions. The BeSiON polytypoids that were detected in the samples are listed in Table 1.2 with their respective Ramsdell symbols [60].

Fig. 1.36 Isothermal section of Si_3N_4–$(Be_3N_2)_2$–$(BeO)_6$–$(SiO_2)_3$ system at 1780 °C. Reprinted with permission of The American Ceramic Society

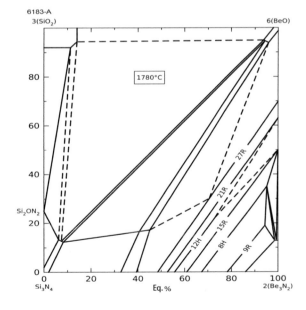

Table 1.2 Compositions of BeSiON polytypoids [60]

Notation	End members	(Si + Be)/ (O + N)	Formula	Homogeneity range (x values)
	Si$_3$N$_4$	3/4	Si$_{3-x}$Be$_x$O$_{2x}$N$_{4-2x}$	0–0.24
	BeSiN$_2$	1/1	Be$_{1+x}$Si$_{1-x}$O$_{2x}$N$_{2-2}$	0–0.165
27R	Be$_6$Si$_3$N$_8$	9/8	Be$_{6+x}$Si$_{3-x}$O$_{2x}$N$_{8-2x}$	0–3
21R	Be$_{11}$Si$_5$N$_{14}$	8/7	Be$_{11+x}$Si$_{5-x}$O$_{2x}$N$_{14}$ $_{-2x}$	0–5
12H	Be$_5$Si$_2$N$_6$	7/6	Be$_{5+x}$Si$_{2-x}$O$_{2x}$N$_{6-2x}$	0–0.95
15R	Be$_9$Si$_3$N$_{10}$	6/5	Be$_{9+x}$Si$_{3-x}$O$_{2x}$N$_{10-2x}$	0–3
8H	Be$_4$SiN$_4$	5/4	Be$_{4+x}$Si$_{1-x}$O$_{2x}$N$_{4-2x}$	0–0.8
9R	Be$_7$SiN$_6$	4/3	Be$_{7+x}$Si$_{1-x}$O$_{2x}$N$_{6-2x}$	0–0.63
	Be$_3$N$_2$	3/2		

References

1. Weiss J, Lukas HL, Lorenz J et al (1981) Calculation of heterogeneous phase equilibria in oxide-nitride systems: I. The quaternary system C–Si–N–O. CALPHAD: Comput Coupling Phase Diagr Thermochem 5(2):125–140
2. Lukas HL, Weiss J, Henig ET (1982) Strategies for the calculation of phase diagrams. CALPHAD: Comput Coupling Phase Diagr Thermochem 6(3):229–251
3. Stull DR, Prophet H (1971) JANAF thermochemical tables, 2nd edn. Natl. Stand. Ref. Data Ser. (U. S., Natl. Bur. Stand.). Rep. No. NSRDS-NBS 37, National Bureau of Standards, U.S. Department of Commerce, Washington, D.C, 1141
4. Hultgren R, Desai PD, Hawkins DT et al (1973) Selected values of the thermodynamic properties of the elements. American Society of Metals, Metals Park, pp 465–471
5. Richter HJ, Herrmann M, Hermel W (1991) Calculation of heterogeneous phase equilibria in the system Si–Mg–N–O. J. Eur Ceram Soc 7(1):3–9
6. Huang ZK, Tien TY (1996) Solid-liquid reaction in the Si$_3$N$_4$–A1N–Y$_2$O$_3$ system under 1 MPa of nitrogen. J Am Ceram Soc 79(6):1717–1719
7. Huang ZK, Tien TY (1994) Solid-liquid reaction in the system Si$_3$N$_4$–Y$_3$Al$_5$O$_{12}$–Y$_2$Si$_2$O$_7$ under 1 MPa of nitrogen. J Am Ceram Soc 77(10):2763–2766
8. Mitomo M, Izumi F, Horiuchi S et al (1982) Phase relationships in the system Si$_3$N$_4$–SiO$_2$–La$_2$O$_3$. J Mater Sci 17(8):2359–2364
9. Wu LE, Sun WZ, Chen YH et al (2011) Phase relations in Si–C–N–O–R (R = La, Gd, Y) systems. J Am Ceram Soc 94(12):4453–4458
10. Lange FF (1980) Si$_3$N$_4$–Ce$_2$O$_3$–SiO$_2$ materials–phase relations and strength. Am Ceram Soc Bull 59(2):239–240, 249
11. Lange FF, Singhal SC, Kuznicki RC (1977) Phase relations and stability studies in the Si$_3$N$_4$–SiO$_2$–Y$_2$O$_3$ pseudoternary system. J Am Ceram Soc 60(5–6):249–252
12. Jack KH (1978) Phase relations in the Si$_3$N$_4$–SiO$_2$–Y$_2$O$_3$ system. Int J Mater Res 11:561–578
13. Cao GZ, Huang ZK, Fu XR et al (1985) Phase equilibrium studies in Si$_2$N$_2$O-containing systems: I. phase relations in the Si$_2$N$_2$O–Al$_2$O$_3$–Y$_2$O$_3$ system. Int J High Technol Ceram 1 (2):119–127
14. Cao GZ, Huang ZK, Fu XR et al (1989) Phase relationship in the Si$_3$N$_4$-Y$_2$O$_3$–La$_2$O$_3$ system. China Sci Ser A 32(4):429–433
15. Nash A, Nash P (1987) The Ge–Ni (Germanium-Nickel) system. Bull Alloy Phase Diagr 8 (3):255–264

16. Cheng YB, Thompson DP (1994) Ceramics, powders, corrosion and advanced processing. Trans Mater Res Soc Jpn 14A:895–898
17. Weiss J, Gauckler LJ, Lukas HL et al (1981) Determination of phase equilibria in the system Si–Al–Zr/N–O by experiment and thermodynamic calculation. J Mater Sci 16(11):2997–3005
18. Lu Y, Huai X, Wu L et al (2015) Phase composition of ZrN–Si_3N_4–Y_2O_3 composite material. J Chin Ceram Soc 43(12):1742–1746
19. Lukas HL, Weiss J, Kreig H et al (1982) Phase equilibria in Si_3N_4 and SiC ceramics. High Temp High Press 14(5):607–616
20. Tien TY, Petzow G, Gauckler LJ et al (1983) Phase equilibrium studies in Si_3N_4-metal oxides systems. In: Riley FL (ed) NATO ASI serial, Serial E, Progress nitrogen ceramics. Kulwer Academic Publishers, Dordrecht, pp 89–99
21. Inomata Y, Hasegawa Y, Matsuyama T (1977) Reaction between Si_3N_4 and MgO as a hot pressing aid. J Ceram Soc Jpn 85:29–31
22. Jack KH (1978) The fabrication of dense nitrogen ceramics. Processing of crystalline ceramics. In: Palimour H., III, Davis RF, Hare TM (eds) Plenum Publishing Corp., New York, pp 561–578
23. Lange FF (1978) Phase relations in the system Si_3N_4–SiO_2–MgO and their interrelation with strength and oxidation. J Am Ceram Soc 61(1–2):53–56
24. Lange FF (1979) Eutectic studies in the system Si_3N_4–Si_2N_2O–Mg_2SiO_4. J Am Ceram Soc 62 (11–12):617–619
25. Sun WY, Yan DS, Tien TY (1988) Subsolidus phase relationships in part of the system Si–Al–Ca–N–O. J Chin Ceram Soc 16(2):130–137
26. Oyama Y, Kamigaito O (1971) Solid solubility of some oxides in Si_3N_4. Jpn J Appl Phys 10 (11):1637
27. Oyama Y (1972) Solid solution in the ternary system, Si_3N_4–AlN–Al_2O_3. Jpn J Appl Phys 11 (5):760
28. Oyama Y (1974) Solid solution in the system silicon nitride-aluminum nitride-aluminum oxide. Yogyo Kyokai Shi 82(7):351–357
29. Jack KH, Wilson WI (1972) Ceramics based on the Si–Al–ON and related systems. Nat Phys Sci 238:28–29
30. Jack KH (1973) Solid solubility of alumina in silicon nitride. Trans J Br Ceram Soc 72:376–378
31. Gauckler LJ, Lukas HL, Petzow G (1975) Contribution to the phase diagram Si_3N_4–AlN–Al_2O_3–SiO_2. J Am Ceram Soc 58(7–8):346–347
32. Layden GK (1976) Process development for pressureless sintering of silicon-aluminum-oxygen-nitrogen ceramic components. United Technologies Research Center, East Hartford
33. Naik IK, Gauckler LJ, Tien TY (1978) Solid-liquid equilibria in the system Si_3N_4–AlN–SiO_2–Al_2O_3. J Am Ceram Soc 61(7–8):332–335
34. Land PL, Wimmer JM, Burns RW et al (1978) Compounds and properties of the system Si–Al–O–N. J Am Ceram Soc 61(1–2):56–60
35. Mao H, Selleby M (2007) Thermodynamic reassessment of the Si_3N_4–AlN–Al_2O_3–SiO_2 system—modeling of the SiAlON and liquid phases. CALPHAD: Comput Coupling Phase Diagrams Thermochem 31(2):269–280
36. Jack KH (1976) SiAlONs and related nitrogen ceramics. J Mater Sci 11(6):1135–1158
37. Jack KH (1977) The crystal chemistry of the SiAlONs and related nitrogen ceramics. In: Riler Fl (ed) Crystal chemistry. NATO Advanced Study Institute Series, vol. 23. Noordhoff International Publishing, pp 109–128
38. Gauckler LJ (1976) Equilibrium in the systems Si, Al/N, O and Si, Al, Be/N, O. Dissertation, University of Stuttgart
39. Sun WY, Yan DS, Gao L et al (1995) Subsolidus phase relationships in systems Ln_2O_3–Si_3N_4–AlN–Al_2O_3 (R = Nd, Sm). J Eur Ceram Soc 15(4):349–355
40. Sun WY, Huang ZK, Chen JX (1983) Subsolidus phase relationships in the system Y_2O_3–Al_2O_3:AlN–Si_3N_4. Trans J Br Ceram Soc 82(5):173–175

41. Naik K, Tien TY (1979) Subsolidus phase relations in part of the system Si, Al, Y/N, O. J Am Ceram Soc 62(11–12):642–643

42. Thompson DP (1986) Phase relationships in Y–Si–Al–O–N ceramics. In: Tressler RE, Messing GL, Pantano CG, Newnham RE (eds) Tailoring multiphase and composite ceramics, vol 20. Plenum Publishing Corp., New York, pp 79–91

43. Sun WY, Tien TY, Yen TS (1991) Subsolidus phase relationships in part of the system Si, Al, Y/N, O: the system Si$_3$N$_4$–AlN–YN–Al$_2$O$_3$–Y$_2$O$_3$. J Am Ceram Soc 74(11):2753–2758

44. Thompson DP, Korgul P, Hendry A (1983) The structural characterisation of SiAlON polytypoids. NATO ASI Series. Series E Appl Phys (Prog Nitrogen Ceram) 65:61–74

45. Kolitsch U, Seifert HJ, Ludwig T et al (1999) Phase equilibria and crystal chemistry in the Y$_2$O$_3$–Al$_2$O$_3$–SiO$_2$ system. J Mater Sci 14(2):447–455

46. Huang ZK, Tien TY, Yen TS (1986) Subsolidus phase relationships in Si$_3$N$_4$–AlN-rare-earth oxide systems. J Am Ceram Soc 69(10):C241–C242

47. Sun WY, Tien TY, Yen TS (1991) Solubility limits of α′–SiAlON solid solutions in the system Si, Al, Y/N, O. J Am Ceram Soc 74(10):2547–2550

48. Huang ZK, Greil P, Petzow G (1983) Formation of α-Si$_3$N$_4$ solid solutions in the system Si$_3$N$_4$–AlN–Y$_2$O$_3$. J Am Ceram Soc 66(6):C96–C97

49. Huang ZK, Yan DS, Tien TY (1986) Formation of R-α′-SiAlON and phase relations in the systems Si$_3$N$_4$AlN-R$_2$O$_3$ (R = Nd, Sm, Gd, Dy, Er and Yb). J Inorg Mater (Chinese) 1(1):55–63

50. Zhu WH, Wang PL, Jia YX et al (1994) Formation of (Na)-SiAlON. J Inorg Mater (Chinese) 9(1):65–71

51. Huang ZK, Sun WY, Yan DS (1985) Phase relations of the Si$_3$N$_4$–AlN–CaO system. J Mater Sci Lett 4(3):255–259

52. Kuang SF, Huang ZK, Sun WY et al (1990) Phase relationships in the Li$_2$O–Si$_3$N$_4$–AlN system and the formation of lithium-α′-SiAlON. J Mater Sci Lett 9(1):72–74

53. Kuang SF, Huang ZK, Sun WY et al (1990) Phase relationships in the system MgO-Si$_3$N$_4$–AlN. J Mater Sci Lett 9(1):69–71

54. Tanaka H, Hasegawa Y, Inomata Y (1978) Phase relations in the system Si$_2$ON$_2$–Y$_2$O$_3$–Al$_2$O$_3$ at 1400 °C. J Ceram Soc Jpn 86(8):365–368

55. Cao GZ, Huang ZK, Fu XR et al (1986) Phase equilibrium studies in Si$_2$N$_2$O-containing systems: II. Phase relations in the Si$_2$N$_2$O–Al$_2$O$_3$–La$_2$O$_3$ and Si$_2$N$_2$O–Al$_2$O$_3$–CaO systems. Int J High Technol Ceram 2(2):115–121

56. Cao GZ, Huang ZK, Fu XR et al (1987) Subsolidus phase relations in Si$_2$N$_2$O-Al$_2$O$_3$–La$_2$O$_3$ and Si$_2$N$_2$O–Al$_2$O$_3$–CaO systems. J Inorg Mat (Chinese) 2(1):54–60

57. Cao G, Huang Z, Fu X et al (1985) Phase relations of Y$_2$O$_3$–Al$_2$O$_3$–Si$_2$N$_2$O. China Sci Ser 4:379–383

58. Wang PL, Tu HY, Sun WY et al (1995) Study on the solid solubility of Al in the melilite system R$_2$Si$_{3-x}$Al$_x$O$_3$ + xN$_{4-x}$ with R = Nd, Sm, Gd, Dy and Y. J Eur Ceram Soc 15:689–695

59. Huang ZK, Chen IW (1996) Rare-earth melilite solid solution and its phase relations with neighboring phases. J Am Ceram Soc 79(8):2091–2097

60. Huseby IC, Lukas HL, Petzow G (1975) Phase equilibria in the system Si$_3$N$_4$–SiO$_2$–BeO–Be$_3$N$_2$. J Am Ceram Soc 58(9–10):377–380

Chapter 2
SiC-Dominated Ceramics Systems

Abstract This chapter presents phase diagrams of many SiC–M_xX_y systems. The M_xX_y stand for one or more compounds of oxides or non-oxides, such as rare earth oxides, Si_3N_4, and Al_4C_3. Due to chemical inertness, SiC seldomly reacts with other compounds. However, compatibility of SiC with neighboring phases, especially the phase relations in SiC–Al_2O_3–R_2O_3 and SiC–Si_3N_4–R_2O_3 systems, is of particular importance for compositional design and sintering of SiC ceramics.

2.1 SiC–SiO$_2$

Weiss et al. [1] calculated pseudobinary section of the SiC–SiO_2 system, as shown in Fig. 2.1. This diagram was part of a larger work on calculated phase equilibria in the Si–C–N–O system, which was investigated by the Powder Metallurgy Laboratory (PML) at Max Planck Institute for Metals Research in Germany in early 1980s. Related ternary systems can be seen in ACerS-NIST Phase Equilibria Diagrams PC Database Version 4.0 (or Version 3.4), or be called PED in this book, and PED 8692, 8693, 8697, 8701-8705, 8745, and 8780 may be referred for detail. The calculation process was further described in PED 8697. Calculated equilibria of SiO_2–C system are presented in PED 9070, and it should be helpful to the establishment of stability range of the SiC phase.

Figure 2.1 shows that SiO_2 has a melting point of 1723 °C. Liquid salt (LS) appears at 1723–1800 °C. SiO_2 component, as well as a few of impurity SiO_2 in SiC powder, would be beneficial to liquid phase sintering of SiC at 1800 °C or higher, when SiO_2 and LS react with some sintering aid oxide to form some silicate, like rare earth silicate, as a second phase. This liquid–solid reaction should promote densification of SiC at 1800–1900 °C, a temperature range normally used for sintering SiC ceramics.

Fig. 2.1 Calculated
pseudobinary section of SiC–
SiO$_2$ system. *ML* metallic
liquid; *LS* liquid salt.
Reprinted with permission of
The American Ceramic
Society, www.ceramics.org

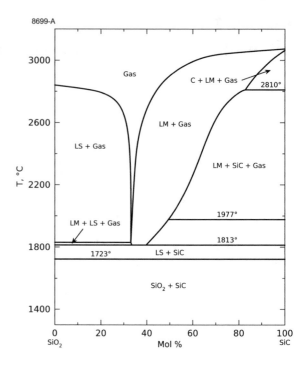

2.2 SiC–Si$_2$ON$_2$

A calculated phase diagram of the SiC–Si$_2$ON$_2$ binary system is shown in Fig. 2.2.
It is a quasibinary vertical section of the system [1]. The phase diagram was
calculated using the program of the PML at Max Planck Institute for Metals
Research in Germany [2]. The data for Si and Si$_2$N$_2$O were picked up from Refs.
[3, 4], respectively, and all other thermodynamic data were taken from the JANAF
table [5]. The calculated results were in good agreement with experimental results
for C–Si–O [6, 7] and C–Si–N [8] systems, which indicates that the calculated
results be a representative for the real equilibrium state. The related ternary systems
are shown in PED 8692, 8693, 8697, 8701-8705, 8745, and 8780.

The phase diagram of the SiC–Si$_2$ON$_2$ system shows that both components are
in equilibrium below subsolidus temperature 1842 °C. Si$_2$ON$_2$ is beneficial to
liquid phase sintering of SiC. It reacts with certain sintering aids such as Y$_2$O$_3$, and
consequently a nitrogen-containing Y-silicate Y$_4$Si$_2$N$_2$O$_7$ (J-phase) is expected to
form. The Y$_4$Si$_2$N$_2$O$_7$ (J-phase) with cuspidine structure has a high melting point
$\sim$ 2000 °C. It would be also in equilibrium with SiC. Ceramics consisted of SiC/
J-phase are expectable to have superior properties, because both Si$_2$ON$_2$ and
Y$_4$Si$_2$N$_2$O$_7$ (J-phase) have been known for excellent oxidation resistance.

Fig. 2.2 Calculated
quasibinary vertical section of
SiC–Si₂ON₂ system.
Reprinted with permission of
The American Ceramic
Society

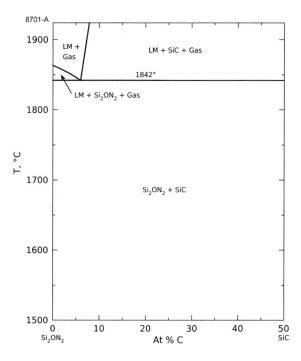

2.3   SiC–Si₃N₄

A calculated pseudobinary section of SiC–Si₃N₄ system [1] is shown in Fig. 2.3.

This calculated diagram was part of a large work on calculated phase equilibria in the Si–C–N–O system, which was investigated by the Powder Metallurgy Laboratory (PML) of Max Planck Institute for Metals Research in Germany. Related ternary systems can be seen in PED 8692, 8697, 8699, 8701-8705, 8745, and 8780. The calculation process was described in detail in PED 8692.

As well known, SiC has high hardness and low toughness, while Si₃N₄ has lower hardness and higher toughness. For this reason, SiC and Si₃N₄ may complement each other in mechanical properties by forming SiC/Si₃N₄ composites.

The calculated phase diagram shows that SiC and Si₃N₄ are in equilibrium below 1877 °C. But both compounds are highly covalent (see Appendix Table A.5) with chemical sluggishness to cause difficulty in sintering. However, by means of reaction of liquid metal (LM), liquid salt (LS), and/or SiO₂ impurity with some oxides added as a sintering aid to form some liquid and/or secondary silicates, SiC/Si₃N₄ composite ceramics can be densified by liquid phase sintering.

Fig. 2.3 Calculated
pseudobinary section of SiC–
Si₃N₄ system. Reprinted with
permission of The American
Ceramic Society

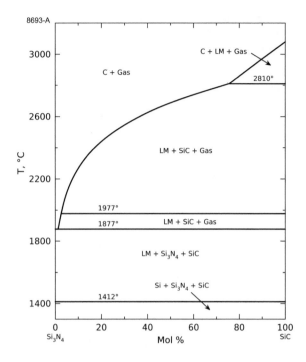

2.4 SiC–Si₃N₄–SiO₂

Calculated isothermal sections [1] of the tetrahedron Si–C–N–O at 1847 and 1807 °C
are shown in Fig. 2.4a, b, respectively. Related ternary systems can be seen in PED
8692, 8693, 8697, 8699, 8701, 8703-8705, 8745, and 8780 in detail. The calculations
were described in PED 8701. The diagrams are skewed triangles as the LS composition was close to, but not exactly, SiO₂. See PED 8745 for schematic location of
the LS composition.

Significance of the calculated isothermal sections for the manufacture of SiC/
Si₃N₄ composite ceramics can be found in the comment in Sect. 2.3.

2.5 SiC–Al₄C₃–Be₂C

An isothermal section of the SiC–Al₄C₃–Be₂C system [9–11] at 1860 °C is shown
in Fig. 2.5.

SiC (99.5%), Al₄C₃ (96.7%), Si (99.7%), Al (99.0%), C (99%), and Be (98.5%)
powders were used as starting materials. Preliminary samples containing Al₄C₃ and
Be₂C were oxidized to introduce Al₄O₄C and BeO, followed by hydrolysis.
Subsequently, elemental components only were used. The powders were mixed
with CCl₄ for 1 h.

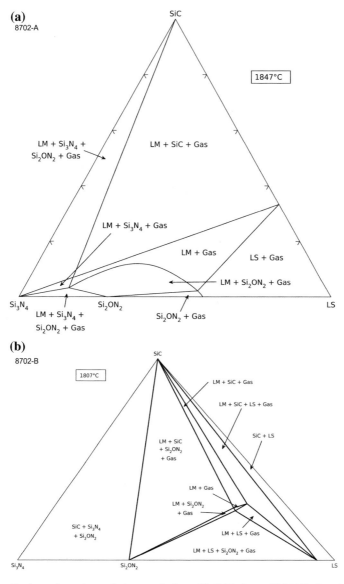

Fig. 2.4 a Isothermal section of the tetrahedron Si–C–N–O at 1847 °C. Reprinted with permission of The American Ceramic Society. **b** Isothermal section of the tetrahedron Si–C–N–O at 1807 °C. Reprinted with permission of The American Ceramic Society

Samples were heated at 1350 °C for a while and then pressured under 28 MPa uniaxial pressure, followed by further heated to 1860 °C and hold for 30 min. Samples were then analyzed by XRD. Seven new phases were reported. Structure and lattice parameter (a, c) of Al$_4$SiC$_4$ and the new phases are shown in Table 2.1.

Fig. 2.5 Isothermal section of SiC–Al$_4$C$_3$–Be$_2$C system at 1860 °C. Reprinted with permission of The American Ceramic Society

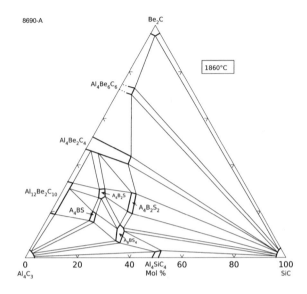

Table 2.1 Crystallographic data of seven newly reported phases and Al$_4$SiC$_4$ in SiC–Al$_4$C$_3$–Be$_2$C system

Compound	Structure	a (nm)	c (nm)
Al$_6$BeC$_5$	Hexagonal	0.3322	1.64
Al$_2$BeC$_2$	Hexagonal	0.3298	2.152
Al$_4$Be$_6$C$_6$	Cubic	0.4611	
Al$_4$SiC$_4$	Hexagonal	0.3281	2.167
Al$_{16}$Be$_4$SiC$_{15}$	Hexagonal	–	–
Al$_{16}$Be$_2$SiC$_{14}$	Hexagonal	–	–
Al$_8$Be$_2$SiC$_8$	Rhombohedra	–	–
Al$_{32}$Be$_2$Si$_4$C$_{29}$	Hexagonal	–	–

2.6 SiC–CrB$_2$

A pseudobinary vertical section of the SiC–CrB$_2$ system at temperatures below 2400 °C is shown in Fig. 2.6, according to Ordan'yan et al. [12]. Both sintered and melted samples were annealed at different temperatures. According to X-ray phase and metallographic analyses, no reaction in either the sintered or fused samples was observed in the samples annealed at different temperatures.

The liquidus curve was set by taking the highest temperature of several compositions within ±10 mol% compositional error as eutectic temperature, allowing extrapolation of the experimental data to the known melting or decomposition temperatures. The pseudobinary nature of the section arises from the known decomposition of SiC at 2760 °C.

Fig. 2.6 Pseudobinary
vertical section of SiC–CrB$_2$
system at temperatures below
2400 °C. Reprinted with
permission of The American
Ceramic Society

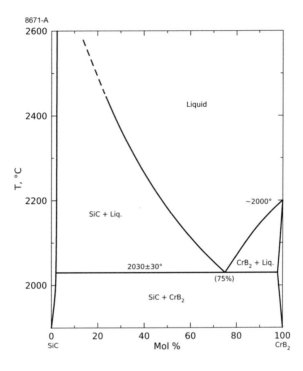

2.7 Al$_4$C$_3$–4(SiC)–B$_4$C

Tentative isothermal section of the Al$_4$C$_3$–4(SiC)–B$_4$C system at 1800 °C is shown in Fig. 2.7, according to Inomata et al. [13]. The Al$_4$C$_3$–SiC system could be seen in PED 8689.

Twenty-nine samples located in the Al$_4$C$_3$ corner were investigated. Neutron activation analysis showed that the Al$_4$C$_3$ contained 3.3 wt% O$_2$. Two new phases Al$_4$Si$_2$C$_5$ and Al$_8$B$_4$C$_7$ were observed. Formula of the Al$_4$Si$_2$C$_5$ compound was confirmed by single-crystal X-ray diffraction data obtained from a small crystal grown at 2000 °C. Large amounts of liquid were observed when these two phases were heated at 1970 and 1780 °C, respectively. Liquid region centered near Al$_8$B$_4$C$_7$ but its extent was not determined. Weight losses of 5 wt% were observed in some samples. Due to selective evaporation of Al and graphite contamination from the dies, the ternary carbides in Fig. 2.7, i.e., Al$_4$Si$_2$C$_5$, Al$_8$B$_4$C$_7$, and Al$_4$SiC$_4$, were assumed as incongruent melting compounds.

Because Al$_4$Si$_2$C$_5$ was observed in ternary mixtures rather than in binary mixtures, it was believed to be unstable at 1800 °C as a pure compound but stabilized by boron. A rhombohedral polymorph of the Al$_4$SiC$_4$ phase was reported by Oscroft et al. [14] in the Al$_4$C$_3$–SiC system, together with two other phases of Al$_4$Si$_3$C$_6$ and Al$_4$Si$_4$C$_7$. Kidwell et al. [15] reported the formation of Al$_8$SiC$_7$, and the phase diagram of Al$_4$C$_3$–SiC system was shown in PED 8689.

Fig. 2.7 Tentative isothermal section of Al_4C_3–4(SiC)–B_4C system at 1800 °C. Reprinted with permission of The American Ceramic Society

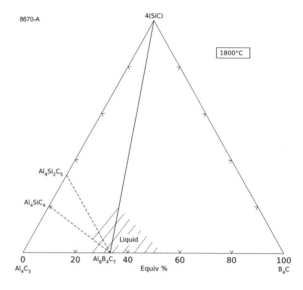

2.8 SiC–SiO₂–Y₂O₃

Phase relationships in SiC–SiO₂–Y₂O₃ system at 1650 °C were calculated by Cupid et al. [16], as shown in Fig. 2.8. The CALPHAD method [17] was used to determine phase equilibria and reactions between the phases. Phase relations were calculated by the software compacts of BINGSS, BINFKT [18], and

Fig. 2.8 Calculated phase diagram in SiC–SiO₂–Y₂O₃ system at 1650 °C. Reprinted with permission of The American Ceramic Society

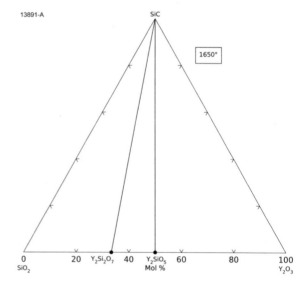

THERMO-CALC [19]. Three 3-phase equilibria at 1650 °C are shown in Fig. 2.8. These equilibria represent possible coating systems based on yttrium silicates and (excess) SiO$_2$ or Y$_2$O$_3$, respectively.

2.9   SiC–Si$_3$N$_4$–R$_2$O$_3$ (R = La, Gd, Y)

Subsolidus phase diagrams for SiC–Si$_3$N$_4$–R$_2$O$_3$ and SiC–Si$_3$N$_4$–SiO$_2$–R$_2$O$_3$ (R = La, Gd, and Y) systems are shown in Fig. 2.9 [20]. La, Gd, and Y were selected to represent light, medium, and heavy rare earth elements, respectively. Twenty-seven samples were prepared to investigate the three ternary systems. Samples were hot pressed at subsolidus temperatures under 30 MPa in Ar or N$_2$ atmosphere for 1–2 h. Samples with weight loss less than 2% were chosen for phase identification, and thus established the phase relations in the SiC–Si$_3$N$_4$–SiO$_2$–R$_2$O$_3$ system.

Solidus phase diagram of SiC–Si$_3$N$_4$–La$_2$O$_3$ system by Wu et al. [20] is shown in Fig. 2.9a. 2Si$_3$N$_4$·La$_2$O$_3$ (2:1 phase, monoclinic) and 2La$_2$O$_3$·Si$_2$ON$_2$ (J-phase, monoclinic cuspidine) are formed between Si$_3$N$_4$ and La$_2$O$_3$. The presence of Si$_2$ON$_2$ in J-phase indicates that the trace SiO$_2$ impurity took part in reactions. SiC coexists with each of them to form the tie lines. Subsolidus phase diagram of SiC–Si$_3$N$_4$–R$_2$O$_3$ (R = Gd, Y) systems [20, 21] is shown in Fig. 2.9b. Si$_3$N$_4$·R$_2$O$_3$ (M-phase, tetragonal Melilite) and 2R$_2$O$_3$·Si$_2$N$_2$O (J-phase, monoclinic cuspidine) are formed both in Gd$_2$O$_3$ and Y$_2$O$_3$ containing systems. SiC coexisted with each of them to establish tie lines. In order to investigate the possible influence of SiO$_2$ impurity in SiC and Si$_3$N$_4$ powders on phase relations, the three ternary systems were extended into quaternary systems [20, 21], as shown in Fig. 2.9c–e. SiC coexisted with the silicates and nitrogen-containing silicates in the Si$_3$N$_4$–SiO$_2$–R$_2$O$_3$ ternary system to establish tie lines linking SiC and each of them.

2.10   SiC–Al$_2$O$_3$–SiO$_2$–Pr$_2$O$_3$

SiC, Al$_2$O$_3$, SiO$_2$, and Pr$_6$O$_{11}$ were used as raw materials for the experimental research. Because Pr$_6$O$_{11}$ releases lattice oxygen at high temperatures, SiC may be oxidized by Pr$_6$O$_{11}$. Therefore, Pr$_2$O$_3$ was prepared by heating Pr$_6$O$_{11}$ at 1200 °C for 2 h in flowing Ar. Samples were prepared in a graphite furnace at 1250–1750 °C for 2–4 h under Ar atmosphere to investigate the phase relations.

Two praseodymium aluminates, namely PrAlO$_3$ (PrAP-type) and PrAl$_{11}$O$_{18}$ [β(Pr)–Al$_2$O$_3$ type], are formed in the Al$_2$O$_3$–Pr$_2$O$_3$ system. Three praseodymium silicates, namely Pr$_2$SiO$_5$, Pr$_2$Si$_2$O$_7$, and Pr$_{9.33}$Si$_6$O$_{26}$ [H(Pr)-apatite-type] are formed owing to SiO$_2$ impurity introduced by the SiC powder. SiC coexisted with each of the aluminates and the silicates to establish the phase relations in SiC–Al$_2$O$_3$–Pr$_2$O$_3$, Al$_2$O$_3$–SiO$_2$–Pr$_2$O$_3$, and SiC–Al$_2$O$_3$–Pr$_2$O$_3$–SiO$_2$ systems, as shown in Fig. 2.10a–c, respectively, according to Pan et al. [22].

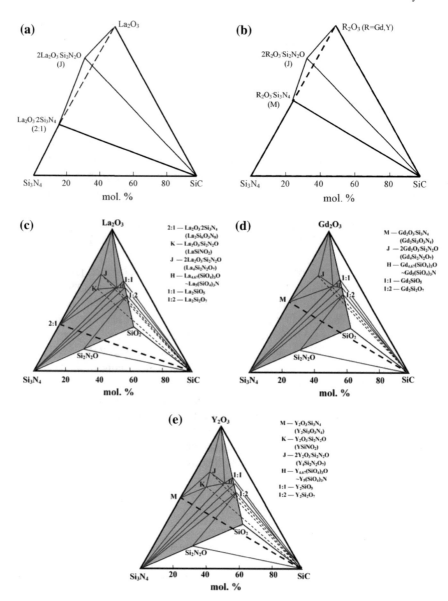

Fig. 2.9 **a** Solidus phase diagram of SiC–Si$_3$N$_4$–La$_2$O$_3$ system. Reprinted from Ref. [20], Copyright 2011, with permission from John Wiley and sons. **b** Solidus phase diagram of SiC–Si$_3$N$_4$–R$_2$O$_3$ (R = Gd, Y) systems. Reprinted from Ref. [20], Copyright 2011, with permission from John Wiley and sons. **c** Tentative phase diagram of SiC–Si$_3$N$_4$–SiO$_2$–La$_2$O$_3$ system. Reprinted from Ref. [20], Copyright 2011, with permission from John Wiley and sons. **d** Tentative phase diagram of SiC–Si$_3$N$_4$–SiO$_2$–Gd$_2$O$_3$ system. Reprinted from Ref. [20], Copyright 2011, with permission from John Wiley and sons. **e** Tentative phase diagram of SiC–Si$_3$N$_4$–SiO$_2$–Y$_2$O$_3$ system. Reprinted from Ref. [20], Copyright 2011, with permission from John Wiley and sons

Fig. 2.10 a Subsolidus phase diagram in SiC–Al$_2$O$_3$–Pr$_2$O$_3$ system. Reprinted from Ref. [22], Copyright 2016, with permission from Göller Verlag. **b** Subsolidus phase diagram in Al$_2$O$_3$–SiO$_2$–Pr$_2$O$_3$ system. Reprinted from Ref. [22], Copyright 2016, with permission from Göller Verlag. **c** Tentative phase diagram in SiC–Al$_2$O$_3$–SiO$_2$–Pr$_2$O$_3$ system. Reprinted from Ref. [22], Copyright 2016, with permission from Göller Verlag

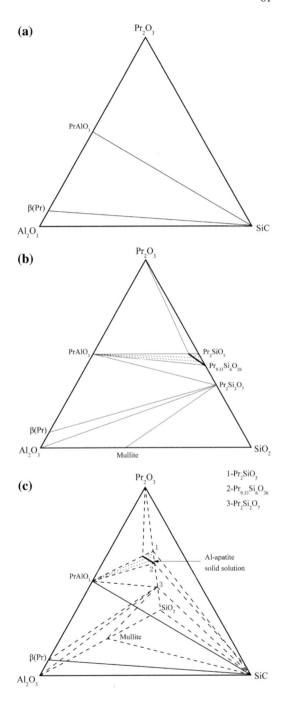

2.11   SiC–Al$_2$O$_3$–SiO$_2$–Nd$_2$O$_3$

Phase relations in SiC–Nd$_2$O$_3$–SiO$_2$, SiC–Al$_2$O$_3$–Nd$_2$O$_3$, and SiC–Al$_2$O$_3$–Nd$_2$O$_3$–SiO$_2$ systems obtained by Ma et al. [23] are shown in Fig. 2.11a–d. The phase relations were determined by using 16 samples fabricated by solid-state reaction at 1650–1720 °C and analyzed by XRD. SiC was compatible with each of the four aluminates, namely Al$_6$Si$_2$O$_{13}$ (mullite), Nd$_2$SiO$_5$, Nd$_{9.33}$[SiO$_4$]$_6$O$_2$(Nd–H-phase) and Nd$_2$Si$_2$O$_7$ at temperature below 1700 °C. NdAlO$_3$ (NdAP) and NdAl$_{11}$O$_{18}$ (Nd–β–Al$_2$O$_3$) were formed in Nd$_2$O$_3$–Al$_2$O$_3$ system. The controversial Nd$_4$Al$_2$O$_9$ (NdAM) phase was formed only at temperature above 1720 °C; it was not detected at temperatures below 1700 °C.

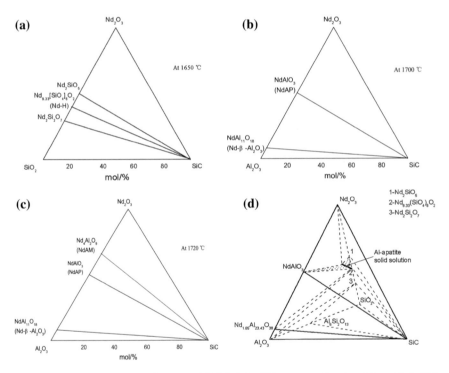

Fig. 2.11 a Subsolidus phase diagram of SiC–Nd$_2$O$_3$–SiO$_2$ system. Reprinted from Ref. [23], Copyright 2016, with permission from Springer Nature. **b** Subsolidus phase diagram of SiC–Al$_2$O$_3$–Nd$_2$O$_3$ system at 1700 °C. Reprinted from Ref. [23], Copyright 2016, with permission from Springer Nature. **c** Subsolidus phase diagram of SiC–Al$_2$O$_3$–Nd$_2$O$_3$ system at 1720 °C. Reprinted from Ref. [23], Copyright 2016, with permission from Springer Nature. **d** Tentative phase diagram of SiC–Al$_2$O$_3$–Nd$_2$O$_3$–SiO$_2$ system. Reprinted, from Ref. [23], Copyright 2016, with permission from Springer Nature

2.12 SiC–Al$_2$O$_3$–SiO$_2$–Gd$_2$O$_3$

Samples were prepared by solid-state reaction through pressureless sintering or hot pressing. SiC, Al$_2$O$_3$, Gd$_2$O$_3$, and SiO$_2$ powders are used as starting materials. Six binary oxides, namely Gd$_4$Al$_2$O$_9$ (GdAM), GdAlO$_3$ (GdAP), Gd$_2$SiO$_5$, Gd$_2$Si$_2$O$_7$, Gd$_{9.33}$(SiO$_4$)$_6$O$_2$ (GdH), and mullite, had formed and SiC was compatible with each of them. Two solid solutions, i.e., Gd$_4$Al$_{2(1-x)}$Si$_{2x}$O$_{9+x}$ (GdAMss) on the GdAM–Gd$_2$SiO$_5$ tie line and the Gd$_{9.33+2x}$(Si$_{1-x}$Al$_x$O$_4$)$_6$O$_2$ (GdHss, Gd-apatite) on the Gd$_{9.33}$ (SiO$_4$)$_6$O$_2$–GdAM tie line, had formed. GdAP could coexist with GdH, Al$_2$O$_3$, and SiC, but it was incompatible with either Gd$_2$Si$_2$O$_7$ or mullite. Accordingly, seven four-phase compatible regions were established in subsolidus phase diagram of the Gd$_2$O$_3$–Al$_2$O$_3$–SiO$_2$–SiC quaternary system by Wei et al. [24], consisting of three ternary systems, namely SiC–Al$_2$O$_3$–Gd$_2$O$_3$, SiC–SiO$_2$–Gd$_2$O$_3$, and Al$_2$O$_3$–SiO$_2$–Gd$_2$O$_3$, as shown in Fig. 2.12a–d.

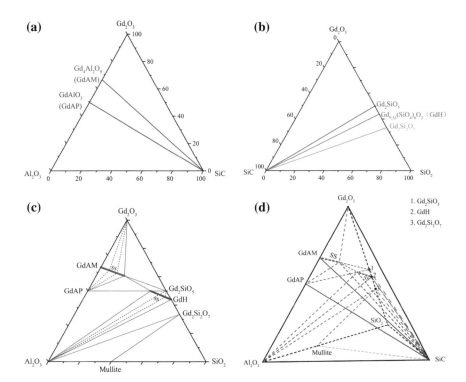

Fig. 2.12 a Subsolidus phase diagram of SiC–Al$_2$O$_3$–Gd$_2$O$_3$ system. Reprinted from Ref. [24], Copyright 2016, with permission from Elsevier. **b** Subsolidus phase diagram of SiC–SiO$_2$–Gd$_2$O$_3$ system. Reprinted from Ref. [24], Copyright 2016, with permission from Elsevier. **c** Subsolidus phase diagram of Al$_2$O$_3$–SiO$_2$–Gd$_2$O$_3$ system. Reprinted from Ref. [24], Copyright 2016, with permission from Elsevier. **d** Tentative phase diagram of SiC–Al$_2$O$_3$–SiO$_2$–Gd$_2$O$_3$ system. Reprinted from Ref. [24], Copyright 2016, with permission from Elsevier

2.13   SiC–Al$_2$O$_3$–SiO$_2$–Yb$_2$O$_3$

Samples were prepared by solid-state reaction at 1400–1750 °C in Ar atmosphere. SiC, Al$_2$O$_3$, Yb$_2$O$_3$, and SiO$_2$ powders were used as starting materials. Formations of Yb$_4$Al$_2$O$_9$ (YbAM), Yb$_3$Al$_5$O$_{12}$ (YbAG), Yb$_2$SiO$_5$, Yb$_2$Si$_2$O$_7$, mullite, and a solid solution of Yb$_4$Al$_{2(1-x)}$Si$_{2x}$O$_{9-x}$(YbAMss) on YbAM–Yb$_2$SiO$_5$ tie line were confirmed by XRD analysis. SiC was compatible with each of these phases, as well as the oxides of Al$_2$O$_3$, Yb$_2$O$_3$, and SiO$_2$, whereas YbAG was compatible with Yb$_2$Si$_2$O$_7$, Al$_2$O$_3$, and SiC, but was not compatible with mullite. Based on these observations, six compatible four-phase regions were established in the SiC–Yb$_2$O$_3$–Al$_2$O$_3$–SiO$_2$ phase diagram, as shown in Fig. 2.13a, b, according to Wei et al. [25].

Fig. 2.13 a Subsolidus phase diagram of SiC–Al$_2$O$_3$–Yb$_2$O$_3$ system. Reprinted from Ref. [25], Copyright 2016, with permission from Elsevier. **b** Tentative phase diagram of SiC–Al$_2$O$_3$–Yb$_2$O$_3$–SiO$_2$ system. Reprinted from Ref. [25], Copyright 2016, with permission from Elsevier

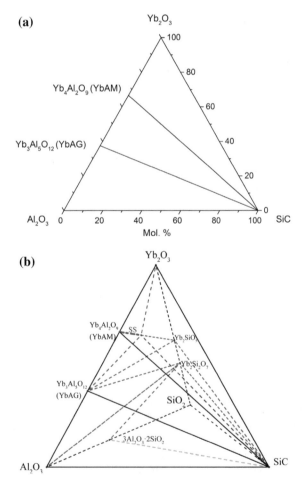

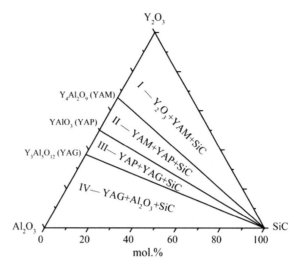

Fig. 2.14 Subsolidus phase diagram of SiC–Al₂O₃–Y₂O₃ system at 1700 °C. Reprinted from Ref. [26], Copyright 2016, with permission from Elsevier

2.14 SiC–Al₂O₃–SiO₂–Y₂O₃

Samples were prepared by solid-state reaction using SiC, Al₂O₃, and Y₂O₃ powders as starting materials. Phase diagram of SiC–Al₂O₃–Y₂O₃ ternary system at 1700 °C was proposed by Jiang et al. [26] according to XRD analysis results, as shown in Fig. 2.14. SiC was compatible with each of the three yttrium aluminates, namely Y₄Al₂O₉ (YAM), YAlO₃ (YAP), and Y₃Al₅O₁₂ (YAG). Stability of YAP phase and possible influence of SiO₂ impurity on phase relations were discussed.

2.15 SiC–SiO₂–Al₂O₃–ZrO₂ and SiC–SiO₂–Al₂O₃–MgO

Samples were prepared by solid-state reaction to investigate phase relations in SiC–SiO₂–Al₂O₃–ZrO₂ and SiC–SiO₂–Al₂O₃–MgO systems. Subsolidus phase diagrams for the two quaternary systems were constructed [27]. SiC coexisted with all the oxides. No ternary compounds were formed while two binary compounds of ZrSiO₄ (azorite) and Al₆Si₂O₁₃ (mullite) were formed in the ZrO₂-containing system. SiC could coexist with the oxides to form three compatible four-phase tetrahedrons.

In the MgO-containing system, besides MgAl₂O₄ (spinel), mullite and binary magnesium silicates of MgSiO₃ and Mg₂SiO₄, two ternary compounds, i.e., Mg₂Al₄Si₅O₁₈ (cordierite) and (Mg₄Al₄)(Al₄Si₂)O₂₀ (emerald), were formed. Emerald was unstable at temperatures below 1450 °C but it was stabilized at temperatures above 1500 °C. Subsolidus phase diagrams of SiC–SiO₂–Al₂O₃–ZrO₂ at 1550 °C (a) and SiC–SiO₂–Al₂O₃–MgO system at 1450 °C (b) and at 1500 °C (c) are shown in Fig. 2.15a–c, respectively.

Fig. 2.15 a Subsolidus phase diagram of SiC–SiO₂–Al₂O₃–ZrO₂ system at 1550 °C. Reprinted from Ref. [27], Copyright 2017, with permission from Springer Nature. **b** Subsolidus phase diagram of SiC–SiO₂–Al₂O₃–MgO system at 1450 °C. Reprinted from Ref. [27], Copyright 2017, with permission from Springer Nature. **c** Subsolidus phase diagram of SiC–SiO₂–Al₂O₃–MgO system at 1500 °C. Reprinted from Ref. [27], Copyright 2017, with permission from Springer Nature

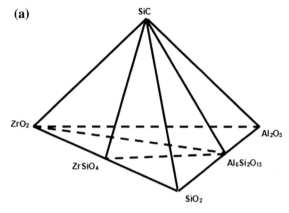

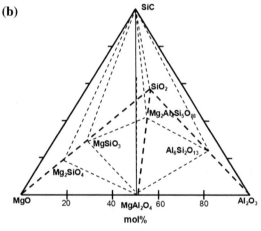

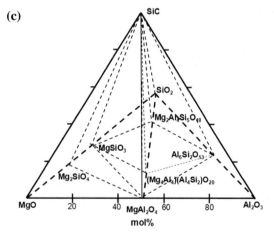

References

1. Weiss J, Lukas HL, Lorenz J et al (1981) Calculation of heterogeneous phase equilibria in oxide-nitride systems: I. The quaternary system C–Si–N–O. CALPHAD Comput Coupling Phase Diagr Thermochem 5(2):125–140
2. Lukas HL, Weiss J, Th Henig E (1982) Strategies for the calculation of phase diagrams. CALPHAD Comput Coupling Phase Diagr Thermochem 6(3):229–251
3. Hultgren R, Desai P, Hawkins D et al (1973) Selected values of the thermodynamic properties of the elements. American Society of Metals, Metals Park, Ohio, pp 465–471
4. Blegen K (1975) Equilibria and kinetics in the systems Si–N and Si–N–O. Special Ceram 6:238–244
5. Stull DR, Prophet H (1971) JANAF thermochemical tables, 2nd edn. National Standard Reference Data System (U.S., Natl. Bur. Stand.), Rep. No. NSRDS-NBS 37, National Bureau of Standards, U.S. Department of Commerce, Washington, D.C., 1141
6. Krivsky WA, Schuhmann R (1961) Derivation of phase diagram for the silicon-oxygen-carbon system. Trans Metall Soc AIME 221(5):898–904
7. Gugel E, Ettmayer P, Schmidt A (1968) Untersuchungen zum System Silizium-Kohlenstoff-Stickstoff. Ber Dtsch Keram Ges 45(8):395–402
8. Rassaerts H, Schmidt A (1966) Thermodynamische Berechnungen im System Silizium-Kohlenstoff-Stickstoff. Planseeber. Pulvermet14(2):110–114
9. Zangvil A, Gauckler LJ, Schneider G et al (1979) TEM studies on $Al_4C_3 \cdot 3Be_2C$. J Mater Sci 14(11):2741–2746
10. Schneider G, Gauckler LJ, Petzow G et al (1979) Phase equilibria in the system Al_4C_3–Be_2C–SiC. J Am Ceram Soc 62(11–12):574–576
11. Schneider G, Gauckler LJ, Petzow G (1979) Phase equilibria in the Si, Al, Be/C, N system. Ceramurgia Int 5(3):101–104
12. Ordan'yan SS, Dmitriev AI, Kapitonova IM (1991) Reaction of SiC with CrB2 Izv. Akad Nauk SSSR, Neorg Mater 27(2):157–159; Inorg Mater (Engl. Transl.) 27(1):134–136
13. Inomata Y, Tanaka H, Inoue HZ et al (1980) Tentative isothermal section of SiC–Al_4C_3–B_4C system at 1800 °C. Yogyo Kyokaishi 88(6):353–355
14. Oscroft RJ, Korgul P, Thompson DP (1989) British ceramics proceedings, complex microstructure, vol 42. Maney Publishing, London, pp 33–47
15. Kidwell BL, Oden LL, McCune RA (1984) $2Al_4C_3SiC$: a new intermediate phase in the Al–Si–C system. J Appl Crystallogr 17(6):481–482
16. Cupid DM, Seifert HJ (2007) Thermodynamic calculations and phase stabilities in the Y–Si–C–O system. J Phase Equilib Diffus 28(1):90–100
17. Saunders N, Miodownik AP (1998) CALPHAD (Calculation of phase diagrams): a comprehensive guide. In: Cahn RW (ed) Pergamon materials series, vol 1. Pergamon, Oxford
18. Lukas HL, Fries SG (1992) Demonstration of the use of "BINGSS" with the Mg–Zn system as example. J Phase Equilibr 13(5):532–541
19. Andersson JO, Helander T, Hoeglund L et al (2002) Thermo-Calc & DICTRA, computational tools for materials science. CALPHAD Comput Coupling Phase Diagr Thermochem 26 (2):273–312
20. Wu LE, Sun WZ, Chen YH et al (2011) phase relations in Si–C–N–O–R (R = La, Gd, Y) systems. J Am Ceram Soc 94(12):4453–4458
21. Sun WZ, Chen YH, Wu LE et al (2011) high temperature phase equilibrium of SiC-based ceramic systems: SiC–Si_3N_4–R_2O_3 (R = Gd, Y) systems. J Mater Sci 46(19):6273–6276
22. Pan WG, Wu LE, Jiang Y et al (2016) Phase relations in the SiC–Al_2O_3–Pr_2O_3 system. J Ceram Sci Technol 7(4):433–440
23. Ma Y, Wu LE, Huang ZK et al (2016) Phase relations in Si–Al–Nd–O–C system. J Phase Equilibr Diffus 37(5):532–539
24. Wei ZB, Jiang Y, Liu LM et al (2016) Phase relations in the Si–Al–Gd –O–C system. Ceram Int 42(2):2605–2610

25. Wei ZB, Jiang Y, Liu LM et al (2016) Phase relations in the Si–Al–Yb–O–C system. J Eur Ceram Soc 36(3):437–441
26. Jiang Y, Wu LE, Wei ZB et al (2016) Phase relations in the SiC–Al$_2$O$_3$–Y$_2$O$_3$ system. Mater Lett 165:26–28
27. Sun WZ, Cheng JG, Huang ZK et al (2017) Phase Relations in Si–Al–M–O–C (M = Zr, Mg) Systems. J Phase Equilibr Diffus 38(1):30–36

Chapter 3
AlN-Based Ceramics Systems

Abstract This chapter includes a series of phase diagrams of AlN-based non-oxide systems. The phase diagrams are collected from that of several systems, i.e., AlN–Al_2O_3, AlN–SiC, AlN–Al_2OC–SiC, and Al_4C_3–AlN–SiC. They are constructed by solid phases of quaternary AlN–Al_4C_3–Al_2O_3–SiC system, in which the AlN–Al_2OC–SiC system was revealed to be a phase diagram of solid solution with tetrahedral [MX_4] wurtzite structure (see Appendix Table A.5). Especially, the phase relations of ternary Al_4C_3–AlN–SiC system have attracted much attention. In addition, AlN–SiC–R_2O_3 (R = light rare earths) systems exist $R_2(Al_{1-x}Si_x)$ $O_3(N_{1-x}C_x)$ solid solution with K_2NiF_4 structure. All these phase diagrams of AlN-based non-oxide ceramics systems are very important for crystal chemistry and physical chemistry.

3.1 Al–Si–N–C–O (AlN–Al_4C_3–Al_2O_3–Si_3N_4–SiC–SiO_2)

The FACT group (Facility for the Analysis of Chemical Thermodynamic) developed the FTOxCN thermodynamic database to calculate phase equilibrium of the Al–(Si–Ca–Mg–Fe–Na)–C–O–N–S system at very high temperatures [1]. The oxycarbonitride phases in the Al–Si–O–C–N (AlN–Al_4C_3–Al_2O_3–Si_3N_4–SiC–SiO_2) system are schematically shown in Fig. 3.1 [1]. All FTOxCN compounds in Fig. 3.1 are stoichiometric oxycarbonitride solid. Al_4C_3·AlN–Al_4C_3·SiC form a continuous solid solution [phase name Al_4C_3(AlN–SiC)] with a primitive hexagonal structure. The solid solution is interrupted in the central composition region by the formation of stoichiometric phase of $2Al_4C_3$·AlN·SiC. One more ternary compound of Al_4C_3·AlN·SiC exists in the system.

Figure 3.1 can be considered as two parts. The first part is the Si_3N_4–AlN–Al_2O_3–SiO_2–SiC quinary system, including phase diagrams of the SiAlONs and the SiC-containing systems (refer to Chaps. 1 and 2), which is the back part of the

---

The original version of this chapter was revised: Belated corrections have been incorporated. The erratum to this chapter is available at https://doi.org/10.1007/978-981-13-0463-7_5

© Springer Nature Singapore Pte Ltd. 2018
Z. Huang and L. Wu, *Phase Equilibria Diagrams of High Temperature Non-oxide Ceramics*, https://doi.org/10.1007/978-981-13-0463-7_3

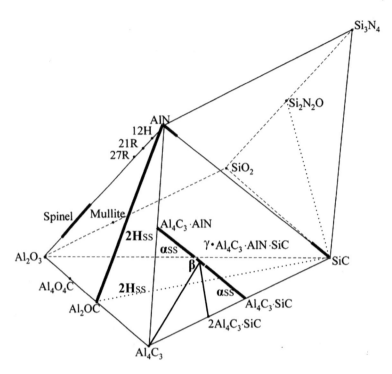

Fig. 3.1 Solid phases in the AlN–Al$_4$C$_3$–Al$_2$O$_3$–Si$_3$N$_4$–SiC–SiO$_2$ system. Reprinted from Ref. [1], with The FACT FTOxCN oxycarbonitride database, from Google Search and Wikipedia

tri-prism. The second part is the AlN–SiC–Al$_4$C$_3$–Al$_2$O$_3$ quaternary system. The calculated and experimental data related to this particular quaternary system will be discussed in the following sections.

3.2   AlN–Al$_2$O$_3$

The experimental phase diagram of the binary system AlN–Al$_2$O$_3$ is shown in Fig. 3.2a [2, 3]. In this system, AlON $\approx$ Al$_{23}$O$_{27}$N$_5$ ss; $\Phi' \approx$ Al$_{22}$O$_{30}$N$_2$ ss, 27R, 21R, and 12H are polytypoids of AlN [4]. γ–Al$_2$O$_3$ (defect cubic spinel, particle size 1.1 μm) and AlN (particle size 14 μm) were milled with alcohol for 24 h, formed under 175 MPa by cold isostatic pressing, and heated at 1200 °C in N$_2$ for 24 h. The heat-treated sample was placed in a BN crucible and was sintered at high temperatures for 1 h. The phase diagram may be better considered as an isobaric section of the binary pressure–temperature–composition (P–T–X) map under $\sim$ 105 Pa N$_2$ pressure.

Figure 3.2b shows a calculated temperature–composition (T–X) phase diagram of AlN–Al$_2$O$_3$ system [5]. In this figure, Spl = spinel composition of AlN·Al2O3-based solid solution, 27R = AlN-rich phase [6], E = eutectic point, and G = gas phase. The compound energy model [5] was applied to calculate the AlN–Al$_2$O$_3$

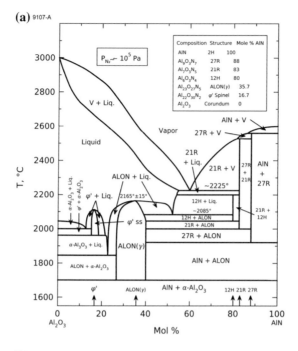

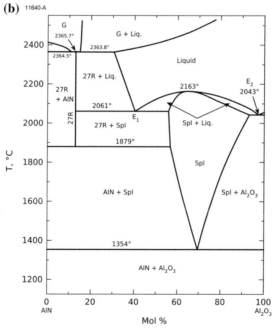

Fig. 3.2 a Experimental phase diagram of AlN–Al₂O₃. Reprinted with permission of The American Ceramic Society. www.ceramics.org. **b** Calculated phase diagram of AlN–Al₂O₃. Reprinted with permission of The American Ceramic Society. **c** Calculated T–X-phase diagram of AlN–Al₂O₃. Reprinted with permission of The American Ceramic Society

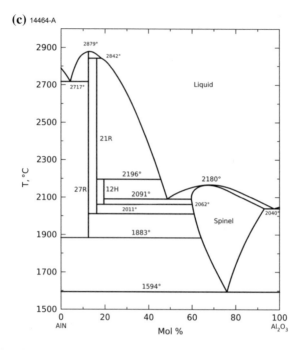

Fig. 3.2 (continued)

system. The stoichiometric $AlN \cdot Al_2O_3$ spinel may be described in the sublattice (or site) terms with the formula $(Al^{3+})_1(Al^{3+})_2(O^{2-},N^{3-})_4$ [5]. This site distribution was found by calculating the minimum Gibbs energy. The main features of the schematic $AlN–Al_2O_3$ phase diagram are shown in the diagram in accordance with Hillert [6]. Reliability of the diagrams has yet to be confirmed by experimental study.

A calculated T–X-phase diagram is shown in Fig. 3.2c [7]. This phase diagram was calculated as part of a thermodynamic assessment of the $SiO_2–Al_2O_3–Si_3N_4–AlN$ section in the system of Si–Al–O–N. General agreement was achieved between calculated results and experimental data.

3.3 $Al_2O_3–Al_4C_3$

Figure 3.3a, b shows the stable diagram (a) and metastable diagram in rapid cooling (b) of the $Al_2O_3–Al_4C_3$ system [8]. Three compositions, containing 10, 20, and 30 mol% of Al_4C_3, were prepared from commercial $\alpha–Al_2O_3$ (corundum) and Al_4C_3. The research has been done by means of XRD, DTA, and microscopy to elucidate the phase relations. It was found that the heating process as well as the cooling rate greatly affected the phases present. For example, more rapid cooling can result in no Al_4O_4C phase, while the Al_2OC phase may present at lower temperatures.

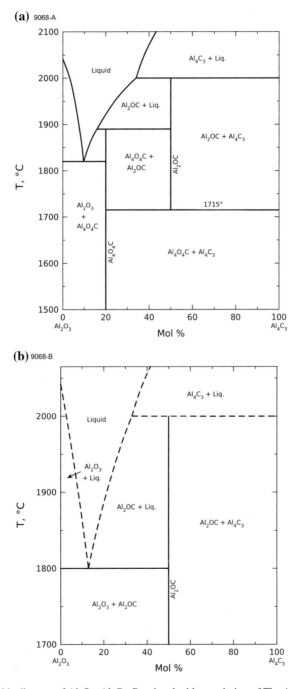

Fig. 3.3 a Stable diagram of Al$_2$O$_3$–Al$_4$C$_3$. Reprinted with permission of The American Ceramic Society. **b** Metastable diagram of Al$_2$O$_3$–Al$_4$C$_3$. Reprinted with permission of The American Ceramic Society. **c** Experimental phase diagram of Al$_2$O$_3$–Al$_4$C$_3$. Reprinted with permission of The American Ceramic Society. **d** Calculated phase diagram of Al$_2$O$_3$–Al$_4$C$_3$. Reprinted from Ref. [10], Copyright 2012, with permission from Springer Nature

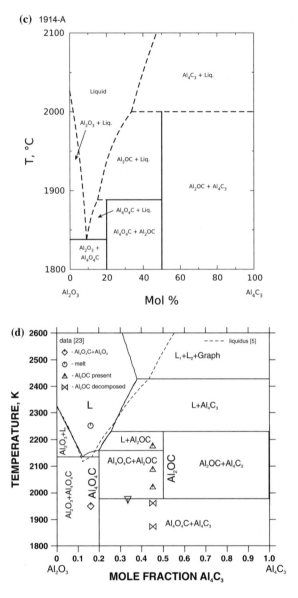

Fig. 3.3 (continued)

Two incongruent melting compounds were determined in Al_2O_3–Al_4C_3 system as shown in Fig. 3.3c [9]. A calculated phase diagram is shown in Fig. 3.3d [10]. Experimental data and liquid phase line of Qiu's [11] were chosen to calculate this binary phase diagram. Two oxycarbides of Al_4O_4C (4:1) and Al_2OC (1:1, 2H-phase) coexist in this system. The calculated enthalpy for the formation of

Al$_2$OC phase was 637.175 kJ/mol and its standard entropy was 33.217 J/(mol K). The calculation showed separation of the liquid phase into two phases, namely a metallic liquid (L1) and an oxycarbide liquid (L2).

3.4 AlN–Al$_4$C$_3$

Figure 3.4 shows the calculated phase diagram of AlN–Al$_4$C$_3$ system. [10]. D. Pavlyuchkov calculated this binary diagram using the optimized parameters of Qiu and Metselaar [11]. One compound Al$_5$C$_3$N exists in this system.

3.5 Al$_2$O$_3$–Al$_4$C$_3$–AlN

This ternary system was first calculated by Qiu [12]. He used experimental data of Lihrmann [13] and description of the liquid ionic model of Mao [7] to process the calculation. Pavlyuchkov et al. [10] treated all solid phase as stoichiometry, and calculated the phase diagram using liquid ionic model. An isothermal section of the Al$_4$C$_3$–Al$_2$O$_3$–AlN system calculated by Pavlyuchkov et al. is shown in Fig. 3.5a, b. This result is in agreement with data of Qiu [12].

3.6 AlN–SiC

Zangvil and Ruh [14] gave a predicted phase diagram of the AlN–SiC binary system as shown in Fig. 3.6. When the temperature is above 2000 °C, continuous solid solution (2H) with wurtzite structure is generated. When heated at temperatures below 2000 °C, the continuous 2H solid solution decomposes to form the miscibility gap consisted of two 2H solid solutions ($\delta_1 + \delta_2$). W. Rafaniello et al. [15] reported stability of the 2H solid solutions in the temperature range between 1700 and 2300 °C, while Pavlyuchkov et al. [10] observed 2H stable as a single phase at temperatures above 2300 °C. This condition indicates separation of the two 2H-phases and formation of the miscibility gap, which was reliably detected in temperature range of 1700–2100 °C. The average temperature value between the abovementioned 2300 °C (2573 K) and 2100 °C (2373 K), i.e., 2473 K, was taken as experimental point determining the maximum of the miscibility gap. The calculated part of the SiC–AlN phase diagram at temperatures below 2500 K was presented by D. Pavlyuchkov et al. as shown in Fig. 3.6b. According to the calculation, separation of the 2H solid solution should start at 2420 K.

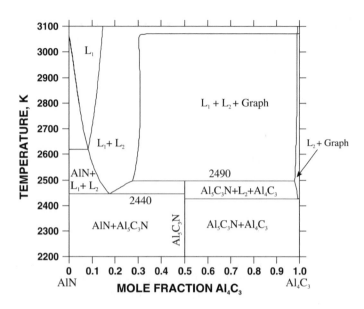

Fig. 3.4 Calculated phase diagram of AlN–Al$_4$C$_3$. Reprinted from Ref. [10], Copyright 2012, with permission from Springer Nature

3.7 AlN–Al$_2$OC–SiC

All the three components in this system have tetrahedral [MX$_4$] wurtzite structure (see Appendix Table A.5) [16]. Continuous solid solutions with the same wurtzite structure, namely xSi$_3$C$_3$·(1 − x)Al$_4$N$_4$, xSi$_3$C$_3$·(1 − x) 2Al$_2$OC, and xAl$_2$OC· (1 − x)2AlN, were formed in the three binary subsystems at high temperatures as shown in Fig. 3.7 [16]. These wurtzite structure continuous solid solutions also possibly decompose into two 2H-phases to form a miscibility gap.

3.8 Al$_4$C$_3$–SiC

Oden et al. [17] provided an experimental pseudobinary phase diagram of the Al$_4$C$_3$–SiC system as shown in Fig. 3.8a. The raw materials of elemental Al, Si, and C were prepared by pre-sintering at 1500 °C for 2 h, grounding, reheating in graphite crucibles to 1600 °C, soaking for 2 h, regrounding, reheating in graphite crucibles to 2000 °C, and soaking again for 2 h, (all under 135 kPa Ar). Samples were produced by heat treatment at a temperature for 1 h, followed by quenching into room temperature oil. Samples for thermal analysis were prepared by hot pressing at 1950 °C for 2 h. Besides the two hexagonal intermediate phases, i.e., Al$_4$SiC$_4$ and Al$_4$Si$_3$C$_6$, three other phases were found. Rhombohedral Al$_4$Si$_2$C$_5$ was found by Inoue et al. [18], hexagonal polytypic Al$_4$Si$_3$C$_6$ was reported by Oscroft

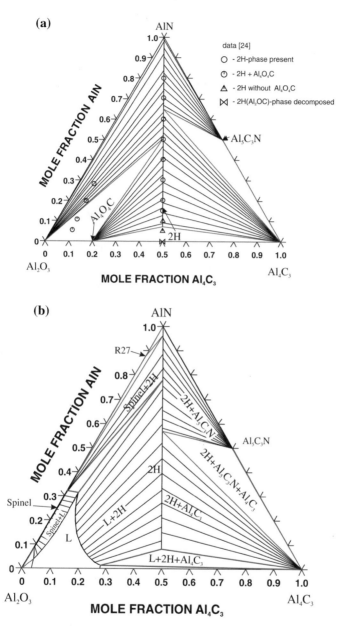

Fig. 3.5 Calculated isothermal section of Al_4C_3–Al_2O_3–AlN system at **a** 1873 K and **b** 2273 K. Reprinted from Ref. [10], Copyright 2012, with permission from Springer Nature

et al. [19], and hexagonal Al_8SiC_7 was observed by Kidwell et al. [20]. Barczak [21] reported that Al_4SiC_4 occurred in two polymorphic forms. A summary of the reported crystallographic data is shown in Table 3.1.

Fig. 3.6 **a** Tentative phase
diagram of AlN–SiC.
Reprinted from Ref. [14],
Copyright 1988, with
permission from John Wiley
and sons, **b** Calculated phase
diagram of SiC–AlN below
2500 K. Reprinted from Ref.
[10], Copyright 2012, with
permission from Springer
Nature

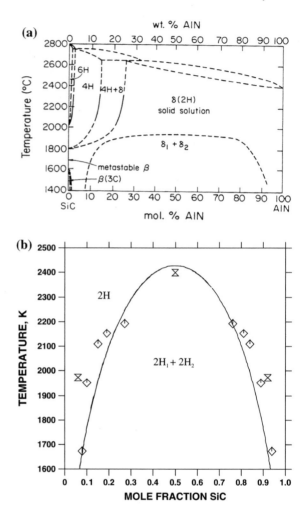

Fig. 3.7 2H solid solution
section of AlN–Al₂OC–SiC.
Reprinted from Google–
Wikipedia Search (http://vip-
digest.narod.ru/archive/html/
1994_001.htm)

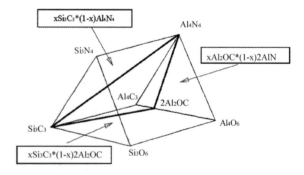

Table 3.1 Phases in Al$_4$C$_3$–SiC system

Phase	System	a (nm)	c (nm)	Reference
α–Al$_4$SiC$_4$	Hexagonal	0.32771	2.1676	[18]
β–Al$_4$SiC$_4$	Rhombohedral	0.3335	3.1326	[19]
Al$_4$Si$_2$C$_5$	Rhombohedral	0.32512	4.01078	[18]
Al$_8$SiC$_7$	Hexagonal	0.33127	1.9242	[20]
Al$_4$Si$_3$C$_6$	Hexagonal	0.2319	3.1784	[19]

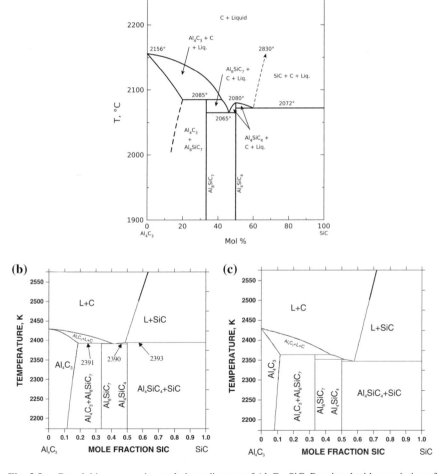

Fig. 3.8 **a** Pseudobinary experimental phase diagram of Al$_4$C$_3$–SiC. Reprinted with permission of The American Ceramic Society. **b** Calculated phase diagram of Al$_4$C$_3$–SiC. Reprinted from Ref. [10], Copyright 2012, with permission from Springer Nature. **c** Calculated phase diagram of Al$_4$C$_3$–SiC. Reprinted from Ref. [22], Copyright 1996, with permission from Elsevier

The system is considered pseudobinary due to the incongruent decomposition of Al_4C_3 at 2156 °C (PED Fig. 8932) and SiC may also decompose below the indicated maximum liquidus temperature.

Figure 3.8b, c shows the calculated phase diagram of Al_4C_3–SiC system once reported by Pavlyuchkov (Fig. 3.8b) [10]. The phase diagram was calculated according to data of Groebner [22] (Fig. 3.8c). What was similar between Fig. 3.8a, b is the existence of an Al_4C_3-based solid solution and formation of two incongruent melting compounds of Al_4SiC_4 (1:1) and Al_8SiC_7 (2:1). However, the incongruent melting points are differed.

3.9 AlN–Al$_4$C$_3$–Si$_3$N$_4$–SiC System

Figure 3.9 presents regular quadrilateral phase diagram of the Si_3N_4–4(AlN)–Al_4C_3–3(SiC) system [23]. Isothermal section at 1760 °C is for portion to the left of the AlN–SiC diagonal and that at 1860 °C is for the portion to the right. A ss, B ss,

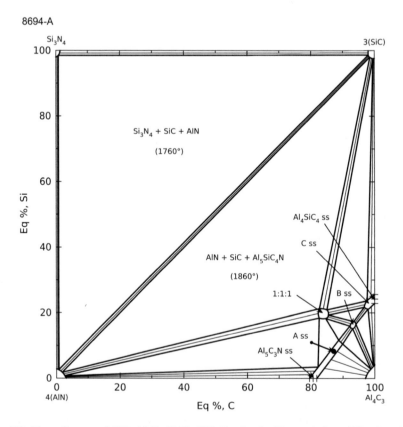

Fig. 3.9 Phase diagram of AlN–Al$_4$C$_3$–Si$_3$N$_4$–SiC. Reprinted with permission of The American Ceramic Society

and C ss are of the general formula Al$_{5-x}$Si$_x$N$_{1-x}$C$_{3+x}$ where x = 0–0.4, 0.5 ± 0.2, and 0.85−1, respectively.

Fifty-eight compositions were prepared from AlN, Si$_3$N$_4$, and C by heating for 45 min at 1760 °C and for 30 min at 1860 °C. Procedures of sample preparation in the N-rich section were described in PED Fig. 8744. Procedures of sample preparation in the C-rich section were described in PED Fig. 8690. No ternary compound was found in the AlN–Si$_3$N$_4$–SiC area. One ternary phase, Al$_5$SiNC$_4$, in the C-rich area has a small homogeneity region. It is a homogeneous area with hexagonal structure having a = 0.325 nm and c = 4.0177 nm. A binary phase, Al$_5$C$_3$N, was found between AlN and Al$_4$C$_3$ [24]. An almost complete solid solution exists between Al$_5$C$_3$N and Al$_4$SiC$_4$. Both A ss and C ss are hexagonal, with a = 0.3277–0.3286 nm and c = 2.1638–2.1755 nm.

3.10 Experimental Phase Diagram of AlN–Al$_4$C$_3$–SiC System

It has been reported that both Al$_4$Si$_2$C$_5$ and Al$_6$C$_3$N$_2$ are rhombohedral. Oden et al. [25] prepared two series of samples with commercially purchased and laboratory-prepared high-purity raw materials to confirm the stability of Al$_4$Si$_2$C$_5$ and Al$_6$C$_3$N$_2$. One series was heated at 2000 °C and the other was sintered by hot pressing. During preparing samples with starting materials of higher purities in laboratory, the hexagonal Al$_6$C$_3$N$_2$ was identified but there was no evidence for the formation of rhombohedral phase. The rhombohedral phase appeared with the substitution of SiC for AlN in the samples. The only phase found to have the rhombohedral structure was Al$_5$SiC$_4$N. In the samples prepared by the commercial material, the rhombohedral structure was found for the Al$_6$C$_3$N$_2$ composition due to oxygen impurity in the commercial AlN stabilizing the rhombohedral form. The isothermal section at 1860 °C of Al$_4$C$_3$–AlN–SiC system, in agreement with the result of Schneider [23], is presented (Fig. 3.10a), where β = Al$_{9+x}$Si$_{1-x}$C$_{7-x}$N$_{1+x}$ ss, −0.04 < x < 0.04. In fact, a miscibility gap exists between Al$_4$C$_3$·AlN and Al$_4$C$_3$·SiC in the region of Al$_4$C$_3$·0.6AlN·0.4SiC to Al$_4$C$_3$·0.15AlN·0.85SiC. The β-phase becomes stable in the region of Al$_4$C$_3$·0.52AlN·0.48SiC to Al$_4$C$_3$·0.48AlN·0.52SiC.

Thermal analysis was also conducted on the hot-pressed specimens prepared by using the higher purity starting materials. The result showed that maximum decomposition temperature of the pseudoternary compound Al$_4$C$_3$·AlN·SiC on the line connecting Al$_4$C$_3$·2AlN–Al$_4$C$_3$·2SiC is 2265 °C. The starting melting temperature is 2175 °C near AlN (Fig. 3.10b), and is 2070 °C near SiC, respectively. The heating reaction is marked by circle and cooling is marked by triangle, respectively, in the figure. The incongruent melting of 2Al$_4$C$_3$·SiC at 2085 °C and Al$_4$C$_3$·SiC at 2080 °C have been observed. The diagram also shows solubility in the end members of the Al$_4$C$_3$–SiC subsystem.

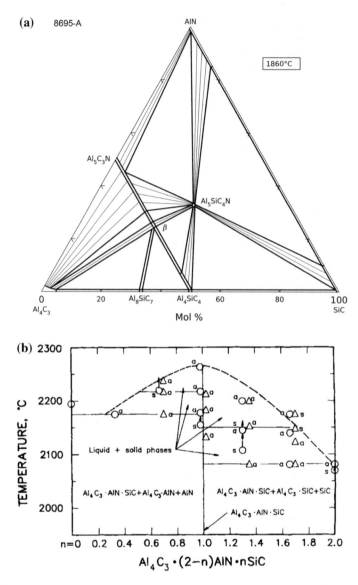

Fig. 3.10 a Isothermal section of SiC–AlN–Al₄C₃ system at 1860 °C. Reprinted with permission of The American Ceramic Society. **b** Thermal analysis of Al₄C₃·(2-n)AlN·nSiC. Reprinted from Ref. [25], Copyright 1990, with permission from John Wiley and sons

3.11 Calculated Phase Diagram of AlN–Al₄C₃–SiC System

The calculated isothermal section at 2133 K by Pavlyuchkov [10] is presented in Fig. 3.11a, together with experimentally determined data by Oden and McCune (Fig. 3.11b) [25]. The thermodynamic parameters of the ternary γ- and β-phases as

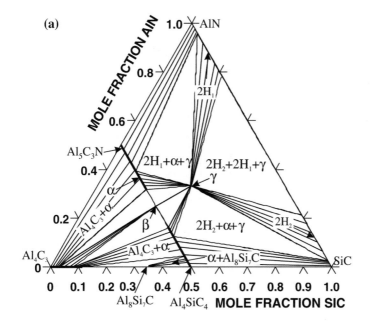

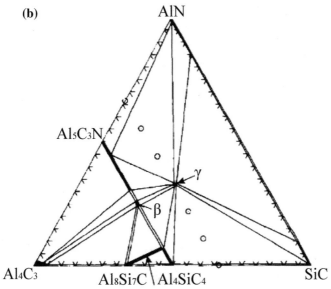

Fig. 3.11 **a** Calculated isothermal section of SiC–AlN–Al$_4$C$_3$ system at 2133 K. Reprinted from Ref. [10], Copyright 2012, with permission from Springer Nature. **b** Experimental phase diagram of SiC–AlN–Al$_4$C$_3$. Reprinted from Ref. [25], Copyright 1990, with permission from John Wiley and sons. In which the line marked with arrow was missed by the authors of Ref. [25]

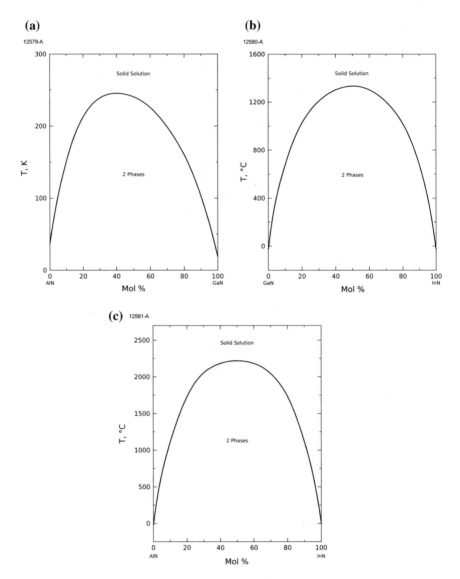

Fig. 3.12 **a** Calculated phase diagram of AlN–GaN binary system. Reprinted with permission of The American Ceramic Society. **b** Calculated phase diagram of GaN–InN binary system. Reprinted with permission of The American Ceramic Society. **c** Calculated phase diagram of AlN–InN binary system. Reprinted with permission of The American Ceramic Society

well as of the α-solid solution were assessed in order to reproduce the experimental results. One can see that phase equilibria of the phase diagram in compositional range between 0 and 50 mol% of Al_4C_3 are well reproduced. On the other hand, the phase equilibria in other part of the diagram (50–100 mol% of Al_4C_3) as well as boundaries of the α-solid solution are different from those presented in Oden and McCune. It should be mentioned that solubility of Al$_5$C$_3$N in Al$_4$SiC$_4$ is higher in Pavlyuchkov's calculations [10] than in those obtained in Oden and McCune's [25].

3.12 AlN–GaN, GaN–InN, and AlN–InN

The AlN–GaN, GaN–InN, and AlN–InN phase diagrams were calculated from first principles by using Vanderbilt-type plane-wave pseudopotentials. Diagrams were calculated with and without the inclusion of the vibrational energy [26]. The difference of the three systems is miscibility gap existing different symmetries, consolute compositions, and temperatures (Fig. 3.12).

3.13 AlN–Eu$_2$O$_3$ and AlN–Nd$_2$O$_3$

Figure 3.13a, b shows the phase diagrams of AlN–Eu$_2$O$_3$ and the AlN–Nd$_2$O$_3$ system, respectively [27]. Starting powders were commercially purchased. Samples were made by mixing the required amounts of the starting powders in an agate mortar under absolute alcohol as a medium for 1.5 h. The mixtures were dried and isostatically pressed under 300 MPa into small cylinders. These cylinders were used for the determination of phase changes during firing and melting behavior. The phase changes were identified by DTA at a heating rate of 10 °C/min, and by visual observation. Hot pressing was also used to promote the solid-state reaction of the samples. Some samples were heat treated after hot pressing at 1400 °C for 24 h in a Pt–Rh wound furnace for the equilibrium studies.

Compound of Eu$_2$AlO$_3$N and Nd$_2$AlO$_3$N (Eu$_2$O$_3$/Nd$_2$O$_3$:AlN in 1:1 mol ratio) was formed in the two systems, respectively [27]. The compound Eu$_2$AlO$_3$N melts incongruently at 1400 °C into a liquid and Eu$_2$O$_3$ with some EuO$_{1-x}$. The compound Nd$_2$AlO$_3$N melts congruently at 1750 °C. For molten samples with compositions near AlN, either big bubbles or large weight loss was observed, indicating the existence of regions of gas phase formation and also proving that the reaction of AlN with Eu$_2$O$_3$ and Nd$_2$O$_3$ promote the decomposition of AlN at high

Table 3.2 Lattice constants of R$_2$AlO$_3$N(1:1) (R = Ce, Pr, Nd, Sm, Eu)

Compound (R$_2$O$_3$/AlN = 1/1)	Tetragonal structure lattice constants (Å)	
	a	c
Ce$_2$AlO$_3$N	3.736	12.72
Pr$_2$AlO$_3$N	3.715	12.60
Nd$_2$AlO$_3$N	3.704	12.505
Sm$_2$AlO$_3$N	3.690	12.371
Eu$_2$AlO$_3$N	3.682	12.38

Table 3.3 Crystallochemical data of BeO, AlN, and SiC compounds

AB compound	BeO	AlN	α–SiC*
Structure	B_4	B_4	B_4
A/B valence ratio	2/2	3/3	4/4
Construction unit	$[BeO_4]$	$[AlO_4]$	$[SiC_4]$
coordination number A/B	4/4	4/4	4/4
bond	Be–O	Al–N	Si–C
Bond length(Å)	1.65	1.87	1.89
Bond ionicity	0.5	0.4	0.2

Note α–SiC: B4, wurtzite structure; β–SiC B2, zinc blende structure

temperatures. The gas phases, as a product of the decomposition of AlN, would be composed of nitrogen and some gaseous oxides of aluminum or of the rare earth. The boundaries of the gas and/or liquid phase regions are indicated by dotted lines because of the difficulties in the measurement. Two compounds are in tetragonal structure; the lattice constants are listed in Table 3.2 [27, 28]. It is obvious that these diagrams are similar to the diagrams of the BeO-containing systems [29] with steep liquidus ties from both end members. It is due to the similar wurtzite structure of AlN and BeO (Table 3.3).

3.14   AlN–Y_2O_3

Figure 3.14a shows the experimental phase diagram of AlN–Y_2O_3 binary system [30]. The preparation of samples and determination of melting behavior are the same explanation for Fig. 3.13. This system shows a single binary system, without formation of compound. It is worth to note that AlN, BeO, and SiC belong to wurtzite compounds with tetrahedron structure, which have relatively lower eutectic point with rare earth oxides [29]. In the central section of AlN–Y_2O_3 system, it contains a wide two "liquid + AlN" phase region above 1730 °C. It can be used as effective sintering additive for densification of nitride ceramics, and many research works had confirmed this conclusion [31]. For example, the sintering behaviors of the AlN(+80 wt%Si_3N_4)–Y_2O_3(+80 wt%Si_3N_4)–La_2O_3(+80 wt%Si_3N_4) system at 1800 °C/3 h/N_2 had been determined as shown in Fig. 3.14b, showing a big dense area with $\geq 98\%$ theoretical density [31]. It is of particular importance for compositional design and manufacture of Si_3N_4 and SiC ceramics.

The calculated phase diagram of the AlN–Y_2O_3 binary system is shown in Fig. 3.14c [32]. The solidus temperature is much higher than that of the experimental phase diagram (Fig. 3.14a), and the temperature and composition of the eutectic point are also different.

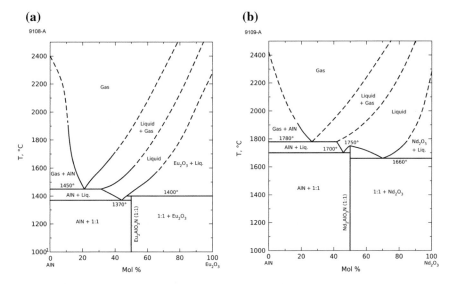

(a)

9108-A

(b)

9109-A

Fig. 3.13 a Phase diagram of AlN–Eu$_2$O$_3$ binary system. Reprinted with permission of The American Ceramic Society. **b** Phase diagram of AlN–Nd$_2$O$_3$ binary system. Reprinted with permission of The American Ceramic Society

3.15 2(Mg₃N₂)–4(AlN)–2(Al₂O₃)–6(MgO)

Solid-state compatibility of this system at 1800 °C is shown in Fig. 3.15 [33]. Spl = spinel (MgAl$_2$O$_4$), Per = periclase (MgO), 21R = a phase of the 21R polytype of the AlN structure, and R = a phase of either the 2Hδ or 27R polytype of the AlN structure [33]. The R-phase is described as having broad and somewhat diffuse X-ray diffraction peaks. Further work on this system can be seen in PED 9111 and 9113.

3.16 AlN–Al₂O₃–Mg₃N₂–MgO

Weiss et al. presented a phase diagram of the AlN–Al$_2$O$_3$–Mg$_3$N$_2$–MgO system as shown in Fig. 3.16 [34], where Q = Mg$_y$Al$_{3-y-(1/3)x}$O$_{(2/3)x}$O$_{3+x+y}$N$_{1-x-y}$,
R = Al$_{(8/3+x/3)}$O$_{(y/3-x/3)}$O$_{4-x}$N$_x$, o = vacancy, and Spl = spinel-type Al$_3$O$_3$N.

The composition at the intersection of the diagonals of the square of the reciprocal salt system prepared from AlN and MgO as well as from Mg$_3$N$_2$ and Al$_2$O$_3$ yielded identical results, even though all compositions on the Mg$_3$N$_2$ side of the line AlN–MgO show a vaporization at 1750 °C (2023 K).

The solid solubility range of the spinel phase was found to correspond to a constant anion sublattice model with a general formula Mg$_y$Al$_{3-y-(1/3)x}$O$_{(2/3)}$$_xO_{3+x+y}N_{1-x-y}$, where $x + y = 1$ with a temperature-dependent solubility limit. The

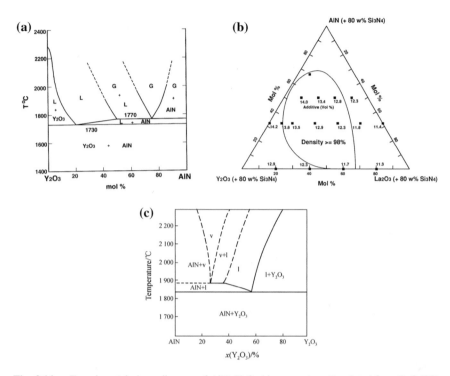

Fig. 3.14 **a** Experimental phase diagram of AlN–Y$_2$O$_3$ binary system. Reprinted from Ref. [30], Copyright 1996, with permission from John Wiley and sons. **b** The sintering behaviors of the AlN (+80 wt%Si$_3$N$_4$)–Y$_2$O$_3$ (+80 wt%Si$_3$N$_4$)–La$_2$O$_3$(+80 wt%Si$_3$N$_4$) system at 1800 °C/3 h/N$_2$. Reprinted from Ref. [31], Copyright 1997, with permission from John Wiley and sons. **c** Calculated phase diagram of AlN–Y$_2$O$_3$ binary system. Reprinted from Ref. [32]

lattice constants measured along the lines of the composition MgAl$_2$O$_4$–Al$_3$O$_3$N and Al$_2$O$_3$–Al$_3$O$_3$N were found to intersect at the composition of the ideal aluminum oxynitride spinel Al$_3$O$_3$N. Specimens of this composition, however, were not found to be single phase at 1750 °C but contained AlN as a second phase. The results found in Weiss et al. work are in good agreement with those found earlier by Jack et al. [33] except for the solubility of the spinel, which was extended over the MgAl$_2$O$_4$–Al$_2$O$_3$N line in work of Jack, and the structure indexing of the "6:1" phase was given as a hexagonal AlN-type structure by Weiss et al. [34].

3.17 2(Mg$_3$N$_2$)–4(AlN)–2(Al$_2$O$_3$)–6(MgO)

Figure 3.17 shows the subsolidus phase diagram of 2(Mg$_3$N$_2$)–4(AlN)–2(Al$_2$O$_3$)–6(MgO) system at 1800 °C [35]. The compositions were prepared in the AlN–Al$_2$O$_3$–MgO triangle region. The samples were hot pressed to avoid excessive weight loss. Phase analysis was performed by X-ray diffraction. HREM lattice imaging was applied to examine the polytypoids formation. Polytypoids were found

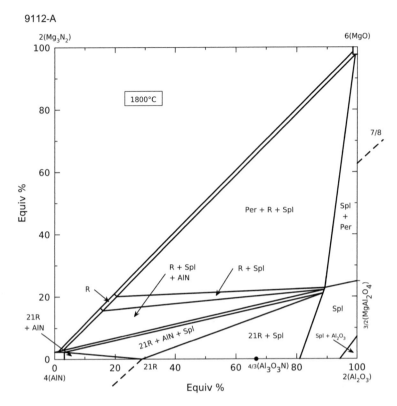

Fig. 3.15 Phase diagram of 2(Mg₃N₂)–4(AlN)–2(Al₂O₃)–6(MgO) system. Reprinted with permission of The American Ceramic Society

to be intergrowths of 9 and 10 layer (27R type), 8 and 9 layer (16H type), and 7 and 8 layer (21R type) blocks when examined by HREM lattice imaging and are typical of low-Mg polytypoids. Magnesium-rich polytypoids of type 14H(7/8) and 8H(8/9) [4] were not found in this study, the non-appearance explained by the authors for the volatility of Mg₃N₂ at the temperature of this study and by recent evidence that these phases were obtained only at 1850 °C [36].

3.18 2(Ca₃N₂)–4(AlN)–2(Al₂O₃)–6(CaO)

The phase diagram of the 2(Ca₃N₂)–4(AlN)–2(Al₂O₃)–6(CaO) system is shown in Fig. 3.18 [37]. Figure 3.18a shows subsolidus phase relationships and (b) for isothermal section at 1700 °C, where a = AlN + CaAl₁₂O₁₉ + AlON ss, b = AlON ss + CaAl₁₂O₁₉ + Al₂O₃, and AlON ss = aluminum oxynitride spinel solid solution.

Compositions were prepared in the triangle region of AlN–CaO–Al₂O₃ due to hydrolysis of Ca₃N₂. Starting materials were AlN, CaO, and Al₂O₃. CaO is prepared by the decomposition of CaCO₃, and Al₂O₃ is prepared by the decomposition

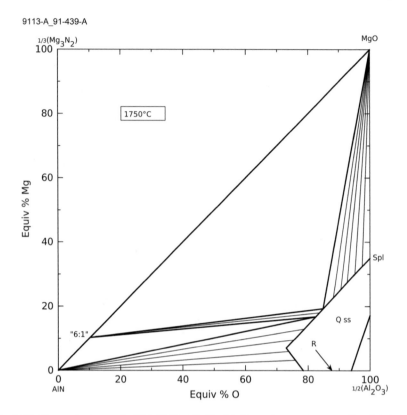

9113-A_91-439-A

Fig. 3.16 Phase diagram of AlN–Al$_2$O$_3$–Mg$_3$N$_2$–MgO quaternary system. Reprinted with permission of The American Ceramic Society

of Ammonia Alum. Mixtures were isostatically pressed at 200 MPa and fired between 1300 and 1750 °C for 1 h under N$_2$. The solid solution range of the Al$_3$ON$_3$ ss spinel solid solution, i.e., Al$_{23}$O$_{27}$N$_5$, was taken from the data of Komeya et al. [38]. The CaCO$_3$-doped AlN (0.3% CaO equivalent) can be sintered to full density at 1800 °C.

3.19   AlN–Al$_2$O$_3$–R$_2$O$_3$ (R = Ce,Pr,Nd,Sm)

Figure 3.19a–e shows the phase diagrams of 4(AlN)–4(RN)–2(Al$_2$O$_3$)–2(R$_2$O$_3$) (R = Ce, Pr, Nd, and Sm) systems [39, 40], in which Fig. 3.19a shows the sub-solidus phase diagram of 4(AlN)–4(NdN)–2(Al$_2$O$_3$)–2(Nd$_2$O$_3$) system, and (b) is an isothermal section at 1700 °C for R = Nd system [39].

The subsolidus phase equilibrium of the AlN–Al$_2$O$_3$–R$_2$O$_3$ (R = Ce, Pr, Nd, and Sm) systems were studied [40]. A new phase with phase composition of RAl$_{11}$O$_{18}$

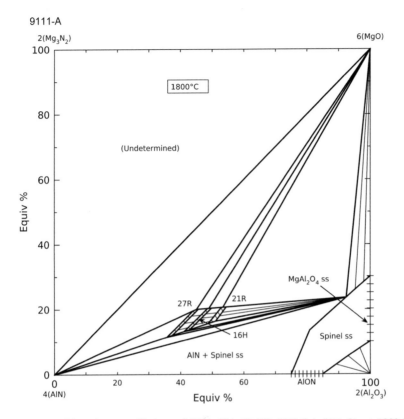

9111-A

Fig. 3.17 Subsolidus phase equilibrium of $2(Mg_3N_2)$–$4(AlN)$–$2(Al_2O_3)$–$6(MgO)$ at 1800 °C. Reprinted with permission of The American Ceramic Society

is analogous to the β-alumina. The investigation of system with other rare earth oxides indicated that $RAl_{11}O_{18}N$ in nitrogenous alumina can be formed from La to Lu (except Pr). The single-phase region was determined as follows: when R = Nd and Sm, nitrogenous alumina only occur at a constitution of $RAl_{11}O_{18}N$; in case of other rare earths, the single phase exists in purity oxides, as $R_2O_3 \cdot 11Al_2O_3$. The lattice constants of $RAl_{11}O_{18}N$ ($a = 5.557$ Å and $c = 22.00$ Å) almost show no change with rare earth elements. The subsolidus phase diagram is quite similar, all of which formed the compounds of $RAl_{12}O_{18}N$, RAP, and R_2AlO_3N, indicating the similar properties of light rare earth elements. A big liquid phase region exists in the Nd_2O_3–Nd_2AlO_3N–$NdAlO_3$ triangle. $NdAl_{11}O_{18}N$ is apparently stable to 1850 °C under N_2 atmosphere [39].

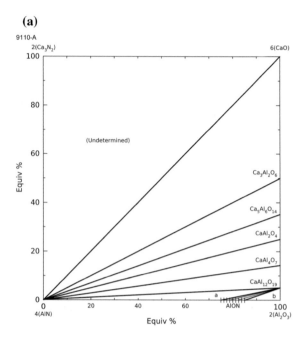

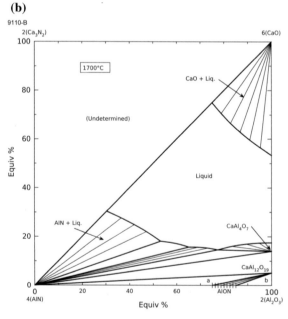

Fig. 3.18 **a** Subsolidus phase diagram of Ca_3N_2–CaO–AlN–Al_2O_3 system. Reprinted with permission of The American Ceramic Society. **b** Isothermal section of Ca_3N_2–CaO–AlN–Al_2O_3 system at 1700 °C. Reprinted with permission of The American Ceramic Society

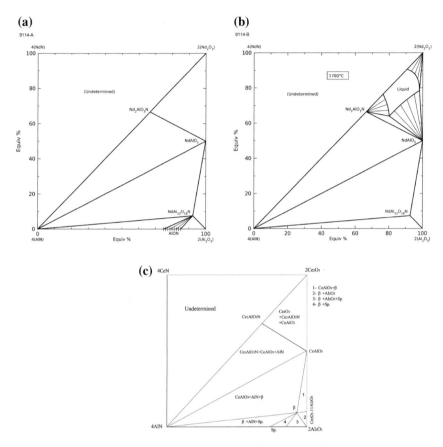

Fig. 3.19 **a** Subsolidus phase diagram of AlN–Al$_2$O$_3$–Nd$_2$O$_3$ system. Reprinted with permission of The American Ceramic Society. **b** Isothermal section AlN–Al$_2$O$_3$–Nd$_2$O$_3$ system at 1700 °C. Reprinted with permission of The American Ceramic Society. **c** Subsolidus phase diagram of AlN–Al$_2$O$_3$–Ce$_2$O$_3$ system. Reprinted from Ref. [40], Copyright 1990, with permission from China Science (Series A). **d** Subsolidus phase diagram of AlN–Al$_2$O$_3$–Pr$_2$O$_3$ system. Reprinted from Ref. [40], Copyright 1990, with permission from China Science (Series A). **e** Subsolidus phase diagram of AlN–Al$_2$O$_3$–Sm$_2$O$_3$ system. Reprinted from Ref. [40], Copyright 1990, with permission from China Science (Series A)

3.20 Nd–Al–Si–O–N Jänecke Prism Phase Diagram

Nd–Al–Si–O–N Jänecke prism phase diagram is shown in Figs. 3.20a–d [41], wherein (a) the phases occurring in the Si$_{6-z}$Al$_z$O$_z$N$_{8-z}$ (0 < z < 4)–"Al$_2$O$_3$:AlN"–Al$_2$O$_3$–Nd$_2$O$_3$–SiO$_2$ region, (b) subsolidus phase equilibria at low Al concentrations and in SiO$_2$-rich corner, (c) compatibility polyhedra, and (d) phase equilibria of the β–SiAlON Si$_{6-z}$Al$_z$O$_z$N$_{8-z}$ (0.8 < z<4) and the U-phase Nd$_3$Si$_{3-x}$Al$_{3+x}$O$_{12+x}$N$_{2-x}$ in the oxygen-rich part.

(d)

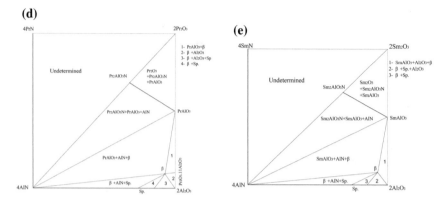

Fig. 3.19 (continued)

Samples for Nd_2O_3–Al_2O_3–SiO_2 system were prepared by hot pressing. For all nitride and oxynitride compositions, a post-sintering heat treatment under nitrogen atmosphere was performed at 1150–1450 °C until no further phase changes were observed. Phase equilibria define the regions of stable coexistence between β–SiAlON $Si_{6-z}Al_zO_zN_{8-z}$ ($0 < z < 4$) and oxide or oxynitride compounds. β–Si_3N_4 coexists with N-melilite ($Nd_2Si_{3-x}Al_xN_{4-x}$, $0 < x < 1$), N-α-wollastonite $NdSi_2ON$, a nitrogen-rich (Al,N)-apatite solid solution, and $Nd_2Si_2O_7$. These compounds are potential grain boundary phases for silicon nitride ceramics. Between $0 \leq z$ 0.8, β–SiAlON with a general formula of $Si_{6-z}Al_zO_zN_{8-z}$ is compatible with N-melilite ($Nd_2Si_{3-x}Al_xN_{4-x}$, $x = 1$), an (Al,N)-apatite of intermediate composition, and $Nd_2Si_2O_7$. The phases in equilibrium with β–SiAlON at compositions of $0.8 \leq z \leq 4$ are $NdAlO_3$, U-phase $Nd_3Si_{3-x}Al_{3+x}O_{12+x}N_{2-x}$ as well as $NdAl_{11+x}O_{18}N_x$ ($x = 1$), and corundum at Al-rich terminal composition ($z = 4$).

3.21 AlN–Be₃N₂–Si₃N₄

Isothermal compatibility of AlN–Be_3N_2–Si_3N_4 system at 1760 °C is shown in Fig. 3.21 [42], wherein A ss = AlN–$BeSiN_2$ solid solution and B ss = Be_4SiN_4 solid solution.

The AlN–Be_3N_2 sample was hot pressed at 1760 °C for 45 min under N_2. Transition of Be_3N_2 from α- to β-phase started at 1600 and completed at 1760 °C, which is consistent with previous work of Rabenau [43]. A solubility limit of 2.5 mol% of Be_3N_2 in AlN was determined, whereas the solubility of AlN in Be_3N_2 could not be similarly derived due to a nonsystematic scattering of the lattice parameters. The lattice parameters of AlN and Si_3N_4 were the same in one- and two-phase samples indicating little solubility between the two phases, which agrees with the work of Gauckler [44]. The solubility of Si_3N_4 in β–Be_3N_2 is less than

(a)

11639-A

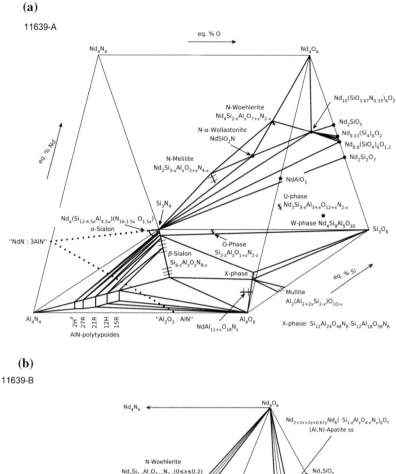

(b)

11639-B

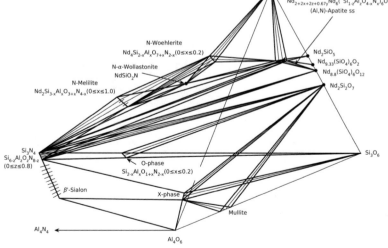

Fig. 3.20 a Jänecke prism phase diagram of $Si_{6-z}Al_zO_zN_{8-z}$ $(0 < z < 4)$–"Al_2O_3:AlN"–Al_2O_3–Nd_2O_3–SiO_2. Reprinted with permission of The American Ceramic Society. **b** Subsolidus phase diagram at low Al concentrations and in SiO_2-rich corner. Reprinted with permission of The American Ceramic Society. **c** Compatibility polyhedral. Reprinted with permission of The American Ceramic Society. **d** Phase equilibria of the β-SiAlON $Si_{6-z}Al_zO_zN_{8-z}$ $(0.8 < z < 4)$ and the U-phase $Nd_3Si_{3-x}Al_{3+x}O_{12+x}N_{2-x}$ in the oxygen-rich part. Reprinted with permission of The American Ceramic Society

(c)

11639-C

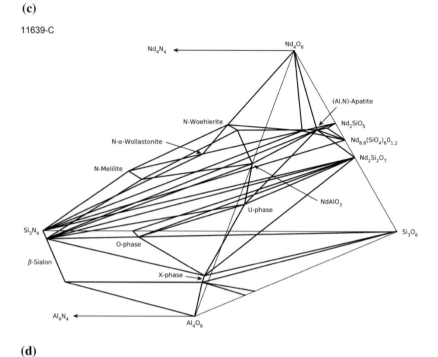

(d)

11639-D

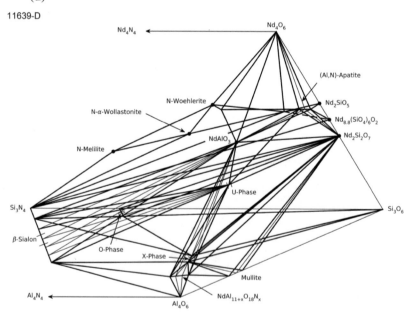

Fig. 3.20 (continued)

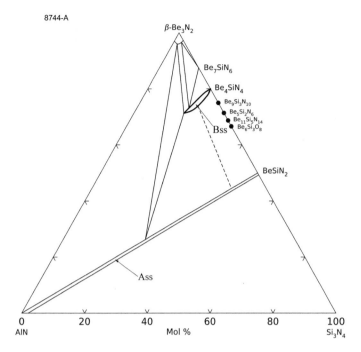

8744-A

Fig. 3.21 Isothermal compatibility of AlN–Be₃N₂–Si₃N₄ system at 1760 °C. Reprinted with permission of The American Ceramic Society

2.5 mol%, which modifies results of Rabenau who determined a solubility of 7 mol %. No detectible Be₃N₂ solubility in Si₃N₄ was found [43]. There is a complete solid solution between AlN and BeSiN₂. Partial solid solution between Be₄SiN₄ and AlN was also found. The other compounds in the β–Be₃N₂–Si₃N₄ binary system have no substantial substitution in the ternary region. The binary Be₃N₂–Si₃N₄ is given in PED Fig. 6183.

3.22 AlN–SiC–R₂O₃ (R: Pr, Nd, Sm, Gd, Yb, Y)

The phase equilibrium in AlN–SiC–Nd₂O₃ system is shown in Fig. 3.22a [45]. The samples were hot pressed under 30 MPa at 1450–1600 °C for 1–2 h in Ar atmosphere. EPMA was used to analyze element components. The results showed that the compound of Nd₂AlO₃N(1:1) exists in the AlN–Nd₂O₃ binary system with melting point at 1750 °C. It belongs to the K₂NiF₄ structure. Si–C in Nd₂O₃:SiC (1:1) can partially substitute for Al–N in Nd₂AlO₃N to form a solid solution with a formula of $Nd_2Al_{1-x}Si_xO_3 N_{1-x}C_x$ ($x = 0–0.5$) (Fig. 3.22a) [45].

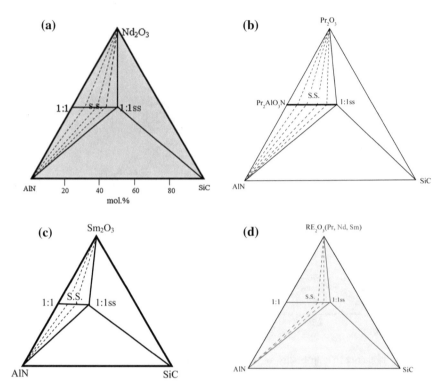

Fig. 3.22 **a** Subsolidus phase diagram of SiC–AlN–Nd₂O₃ ternary system. Reprinted from Ref. [45], Copyright 2013, with permission from Springer Nature. **b** Subsolidus phase diagram of AlN–SiC–Pr₂O₃ system. Reprinted from Ref. [46], Copyright 2017, with permission from Springer Nature. **c** Subsolidus phase diagram of the AlN–SiC–Sm₂O₃ ternary system. Reprinted from Ref. [47], Copyright 2017, with permission from Springer Nature. **d** Overlapped phase diagram of AlN–SiC–Pr₂O₃\Nd₂O₃\Sm₂O₃ system. Reprinted from Ref. [45–47], Copyright 2017, with permission from Springer Nature. **e** Tentative phase diagram of SiC–AlN–Nd₂O₃–Al₂O₃ system. Reprinted from Ref. [45], Copyright 2013, with permission from Springer Nature. **f** Tentative phase diagram of the SiC–AlN–Pr₂O₃–Al₂O₃ ternary system. Reprinted from Ref. [46], Copyright 2017, with permission from Springer Nature. **g** Phase diagram of the AlN–Al₂O₃–Pr₂O₃ ternary system at 1630 °C. Reprinted from Ref. [46], Copyright 2017, with permission from Springer Nature. **h** Phase diagram of AlN–SiC–R₂O₃ (R = Gd,Yb,Y) system. Reprinted from Ref. [45], Copyright 2013, with permission from Springer Nature

Equilibrium phase diagrams in the AlN–SiC–Pr₂O₃ [46] and the AlN–SiC–Sm₂O₃ [47] systems were constructed by similar method (see [45]). Both phase diagrams with the same character are shown in Fig. 3.22b, c [46, 47]. The solid solubility of Si–C in relevant solid solutions was determined as 0.6 and 0.4, respectively, and listed in Table 3.4. With the increase in atomic number of rare earths from Pr to Nd and Sm, i.e., decrease in ionic radius from Pr³⁺ 0.132 nm to Nd³⁺ 0.130 nm and Sm³⁺ 0.127 nm, the solubility x of Si–C in $R_2Al_{1-x}Si_xO_3N_{1-x}C_x$ decreases from 0.6 for Pr to 0.5 for Nd and 0.4 for Sm. The overlapped phase

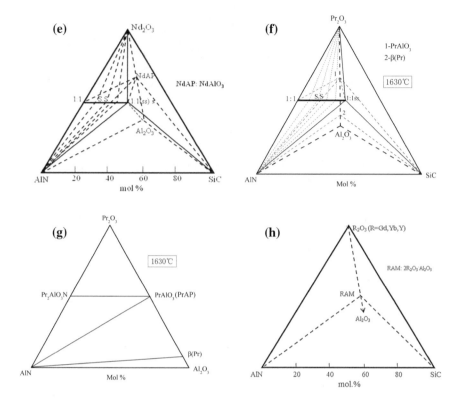

Fig. 3.22 (continued)

Table 3.4 Cation ratios of R$_2$Al$_{1-x}$Si$_x$O$_3$N$_{1-x}$C$_x$ ss phase in AlN:SiC:R$_2$O$_3$ = 2:2:1 samples [45–47]

Phases	R	Si	Al	(Si + Al)/R	X = Si/(Al + Si)
NAS(1:1)ss	26.95	6.33	6.80	0.49	0.48 (~0.5) [45]
SAS(1:1)ss	20.57	5.22	7.05	0.59	0.42 (~0.4) [47]
PAS(1:1)ss	12.19	3.47	2.25	0.47	0.61 (~0.6) [46]

diagram of the AlN–SiC–Pr$_2$O$_3$\Nd$_2$O$_3$\Sm$_2$O$_3$ systems is shown in Fig. 3.22d [45–47].

In addition, a small amount of RAlO$_3$ (RAP, R = Pr,Nd,Sm) was observed due to Al$_2$O$_3$ impurity in AlN powder participating in reaction with rare earth oxide. The tentative phase diagrams of AlN–SiC–Al$_2$O$_3$–Nd$_2$O$_3$,–Pr$_2$O$_3$ systems were established in Fig. 3.22e [45] and (f) [46].

Besides, the reaction of Al$_2$O$_3$ as a member component with AlN–Pr$_2$O$_3$ was also studied [46]. The phase diagram of the AlN–Al$_2$O$_3$–Pr$_2$O$_3$ system at 1630 °C was established as in Fig. 3.22g [46], showing Pr$_2$AlO$_3$N, PrAlO$_3$ (PrAP), and

$\beta(Pr)Al_2O_3$ compounds formation, in which $PrAlO_3$ (PrAP) is in equilibrium with Pr_2AlO_3N and AlN, respectively. The phase diagram of the $AlN-Al_2O_3$ binary system had been reported to form many AlN polytypoids at $\geq 1700\ °C$ (see Fig. 3.2a [2, 3] and Fig. 3.19d [40]); however, no any polytypoid was observed on the join $AlN-Al_2O_3$ in the $AlN-Al_2O_3-Pr_2O_3$ system at $1630\ °C$ [46], due to lower temperature.

With other heavy rare earth elements, neither binary nor ternary compound was found to form in the $AlN-SiC-R_2O_3$ (R = Gd, Yb, Y) systems. The phase diagram was characterized by an empty triangle. Occasional formation of some $R_4Al_2O_9$ (RAM) was due to Al_2O_3 impurity in the AlN starting powder. The tentative diagram of $AlN-SiC-R_2O_3$ (R = Gd, Yb, Y) systems is shown in Fig. 3.22h [45].

References

1. FACT FTOxCN oxycarbonitride database. http://www.crct.polymtl.ca/fact/documentation/FTOxCN/FTOxCN_documentation.htm
2. McCauley JW, Corbin ND (1983) High temperature reactions and microstructures in the Al_2O_3-AlN system. NATO ASI Ser, Ser E Prog Nitrogen Ceram 111–118
3. McCauley JW, Corbin ND (1979) Phase relations and reaction sintering of transparent cubic aluminum oxynitride spinel (ALON). J Am Ceram Soc 62(9–10):476–479
4. Thompson DP, Korgul P, Hendry A (1983) The structural characterisation of SiAlON polytypoids. Progress in nitrogen ceramics. Springer, Netherlands, pp 61–74
5. Barry TI, Dinsdale AT, Gisby JA et al (1992) The compound energy model for ionic solutions with applications to solid oxides. J Phase Equilibr 13(5):459–475
6. Hillert M, Jonsson S (1999) Report No. TRITA-MAC 399. Royal Institute of Technology, Stockholm
7. Mao H, Selleby M (2007) Thermodynamic reassessment of the Si_3N_4-AlN-Al_2O_3-SiO_2 system-modeling of the SiAlON and liquid phases. CALPHAD Comput Coupling Phase Diagr Thermochem 31(2):269–280
8. Larrere Y, Willer B, Lihrmann JM et al (1984) Dagrammes d'équilibre stable et métastable dans le système Al_2O_3-Al_4C_3. Revue internationale des hautes températures et des réfractaires 21(1):3–18
9. Foster LM, Long G, Hunter MS (1956) Reactions between aluminum oxide and carbon the Al_2O_3-Al_4C_3 phase diagram. J Am Ceram Soc 39(1):1–11
10. Pavlyuchkov D, Fabrichnaya O, Herrmann M et al (2012) Thermodynamic assessments of the Al_2O_3-Al_4C_3-AlN and Al_4C_3-AlN–SiC systems. J Phase Equilib Diffus 33(5):357–368
11. Qiu C, Metselaar R (1995) Thermodynamic evaluation of the Al_2O_3-Al_4C_3 system stability of Al-Oxycarbides. Zeitschrift Fur Metallkunde 86(3):198–205
12. Qiu C, Metselaar R (1997) Phase relations in the aluminum carbide-aluminum nitride-aluminum oxide system. J Am Ceram Soc 80(8):2013–2020
13. Lihrmann JM, Tirlocq J, Descamps P et al (1999) Thermodynamics of the Al–C–O system and properties of SiC–AlN–Al_2OC composites. J Eur Ceram Soc 19(16):2781–2787
14. Zangvil A, Ruh R (1988) Phase relationships in the silicon carbide-aluminum nitride system. J Am Ceram Soc 71(10):884–890
15. Rafaniello W, Plichta MR, Virkar AV (1983) Investigation of phase stability in the system SiC-AlN. J Am Ceram Soc 66(4):272–276

16. Schneider A, Schober R (1994) New materials in the AlCON-system. In: Proceedings of International Conference Ceramic Processing Science and Technology, Friedrichshafen (Germany), 11–14, Sept 1994

17. Oden LL, McCune RA (1987) Phase equilibria in the Al–Si–C system. Metall Mater Trans A 18(12):2005–2014

18. Inoue Z, Inomata Y, Tanaka H et al (1980) X-ray crystallographic data on aluminum silicon carbide, α-Al$_4$SiC$_4$ and Al$_4$Si$_2$C$_5$. J Mater Sci 15(3):575–580

19. Oscroft RJ, Korgul P, Thompson DP (1989) British ceramics proceedings. Complex microstructure, vol. 42. Maney Publishing, London, pp 33–47

20. Kidwell BL, Oden LL, McCune RA (1984) 2Al$_4$C$_3$SiC: a new intermediate phase in the Al–Si–C system. J Appl Crystallogr 17(6):481–482

21. Barczak VJ (1961) Optical and X-ray powder diffraction data for Al$_4$SiC$_4$. J Am Ceram Soc 44(6):299

22. Groebner J, Lukas HL, Aldinger F (1996) Thermodynamic calculation of the ternary system Al–Si–C. CALPHAD Comput Coupling Phase Diagr Thermochem 20(2):247–254

23. Schneider G, Gauckler LJ, Petzow G (1979) Phase equilibria in the Si, Al, Be/C, N system. Ceramurgia Int 5(3):101–104

24. Jeffrey GA, Wu V (1966) The structure of the aluminum carbonitrides. II. Acta Crystallogr 20 (4):538–547

25. Oden LL, McCune RA (1990) Contribution to the phase diagram Al$_4$C$_3$-AlN-SiC. J Am Ceram Soc 73(6):1529–1533

26. Burton BP, Walle AVD, Kattner U (2006) First principles phase diagram calculations for the wurtzite-structure systems AlN-GaN, GaN-InN, and AlN-InN. J Appl Phys 100(11):113528/1–113528/6

27. Huang ZK, Yan DS, Tien TY (1990) Compound formation and melting behavior in the AB compound and rare earth oxide systems. J Solid State Chem 85(1):51–55

28. Marchand R (1976) Oxynitrides with potassium nickel(II) tetrafluoride structure. Ln$_2$AlO$_3$N compounds (Ln = lanthanum, neodymium, samarium). Comptes Rendus de l' Academie des Sciences Serie IIc. Chemie 282(7):329–331

29. Yen TS, Kuo CK, Han WL et al (1983) Phase equilibria in the systems rare-earth sesquioxide-beryllium oxide. J Am Ceram Soc 66(12):860–862

30. Huang ZK, Tien TY (1996) Solid-liquid reaction in the Si$_3$N$_4$–AlN–Y$_2$O$_3$ system under 1 MPa of nitrogen. J Am Ceram Soc 79(6):1717–1719

31. Huang ZK, Rosenflanz A, Chen IW (1997) Pressureless sintering of Si$_3$N$_4$ ceramic using AlN and rare-earth oxides. J Am Ceram Soc 80(5):1256–1262

32. Kouhik B (2002) Liquid phase sintering of SiC ceramics with rare earth sesquioxides. Ph.D. Thesis, University of Stuttgart, Stuttgart. http://citeseerx.ist/psu.edu/viewdoc/download?doi=10.1.1.632.4758&rep1&type = pdf

33. Jack KH (1976) SiAlONs and related nitrogen ceramics. J Mater Sci 11(6):1135–1158

34. Weiss J, Greil P, Gauckler LJ (1982) The systme Al–Mg–O–N. J Am Ceram Soc 65(5):C68–C69

35. Sun WY, Ma LT, Yan DS (1990) Phase relationship in Mg–Al–O–N system. Sci Bull (Chinese) 35(14):1189–1192

36. Kuang SF, Huang ZK, Sun WY et al (1990) Phase relationships in the system MgO–Si$_3$N$_4$–AlN. J Mater Sci Lett 9(1):69–71

37. Sun WY, Yen TS (1989) Phase relationships in the system Ca–Al–O–N. Mater Lett 8(5):150–152

38. Komeya K, Tsuge A, Inoue H et al (1982) Effect of CaCO$_3$ addition on the sintering of AlN. J Mater Sci Lett 1(8):325–326

39. Sun WY, Yen TS (1989) Phase relationships in the system Nd–Al–O–N. Mater Lett 8 (5):145–149

40. Sun WY., Chen JX, Jia YX et al (1990) Phase relationships in the system R$_2$O$_3$–AlN–Al$_2$O$_3$ (R = Ce,Pr,Nd,Sm). China Sci A(9):990–998

41. Kaiser A, Telle R, Herrmann M et al (2001) Subsolidus phase relationships of the β-SiAlON solid solution in the oxygen-rich part of the Nd–Si–Al–ON system. Zeitschrift für Metallkunde 92(10):1163–1169

42. Schneider G, Gauckler LJ, Petzow G (1980) Phase equilibria in the system AlN–Si$_3$N$_4$–Be$_3$N$_2$. J Am Ceram Soc 63(1–2):32–35
43. Rabenau A, Eckerlin P (1960) Special ceramics. In: Popper P (ed) Academic Press, vol 1. New York, pp 136–143
44. Gauckler LJ (1975) Equilibrium in the systems Si, Al/N, O and Si, Al, Be/N, O. Ph.D. Dissertation, University of Stuttgart, Stuttgart, Germany
45. Chen YH, Sun WZ, Wu LE et al (2013) Phase relations in SiC–AlN–R$_2$O$_3$ (R = Nd, Gd, Yb, Y) systems. J Phase Equilibr Diffus 34(1):3–8
46. Pan WG, Wu LE, Jiang Y et al (2017) Solid solution and phase relations in SiC–AlN–Pr2O3 system. J Phase Equilibr Diffus 38(5):676–683
47. Jiang Y, Wu LE, Huang ZK (2014) Solid-solution of nitrogen-containing rare earth aluminates R$_2$AlO$_3$N (R = Nd and Sm). Ceram Eng Sci Proc 34(2):95–99

Chapter 4
Ultrahigh-Temperature Ceramics (UHTCs) Systems

Abstract Phase diagrams of diborides (MB_2), monocarbides (MC), and mononitrides (MN) of transitive metals in IVB or VB groups, such as Ti, Zr, Hf, Nb, and Ta, most of which belong to the Ultrahigh-Temperature Ceramic (UHTC) family, are collected in this chapter. Phase diagrams of UHTCs are scarce relative to common oxides, carbides, or nitrides. High melting point and high hardness cause difficult to experiment the phase diagram. Therefore, element diagrams and calculated diagrams in open publications are also included.

4.1 Introduction

Most of the phase diagrams in this chapter are taken from ACerS-NIST Phase Equilibria Diagrams PC Database Version 4.0 (or Version 3.4), or be called PED in this book.

Ultrahigh-temperature ceramics are those with melting points higher than 3000 °C and are aimed to perform at temperatures over 2000 °C. Generally, materials having potential for ultrahigh-temperature applications are consisted of about 19 substances; they are elementary substances C, W, Re, Os, Ta, and compound ThO_2 and non-oxides compounds, i.e., diborides, monocarbides, and mononitrides of transition metals of IVB and VB group in the periodic table (see in Appendix Table A.6), especially Ti, Zr, Hf, Nb, and Ta. Because most those substances and compounds are chemically stable, it is usually difficult for reactions involving them to reach equilibrium, even at high temperatures. Therefore, quite a few phase diagrams of the UHTC systems are based on calculation, in addition to experimental phase diagrams. Most phase diagrams of the UHTC systems were published in the 1960s and 70s by the Air Force Research Laboratory and the former Soviet Union (Kiev) Powder Metallurgy Laboratory. In order to construct the phase diagrams, they first determined ternary phase diagrams of the element–element systems. For example, IVB transition metals (M^{IV}) or their molten alloys were used to

---

The original version of this chapter was revised: Belated corrections have been incorporated. The erratum to this chapter is available at https://doi.org/10.1007/978-981-13-0463-7_5

produce binary or ternary compounds, such as TiC and ZrB_2, or solid solution, such as (Zr, Hf)C ss, by M^{IV} reaction with carbon or boron (M^{IV}–C, –B). Then, phase relationship cross sections between two phases were measured, and binary phase diagrams were accordingly constructed. Ternary phase diagrams were drawn by further addition of other components to the binary phase systems.

4.2 TiB_2–$TiC_{0.9}$ and TiB_2–B_4C

The system is taken from the Ti–B–C element system [1]. It is part of a larger study about borides, carbides, and silicides of the transition metals. The specimens were prepared from Ti powder, graphite, and B powder. The TiB_2 was prepared from the reaction between 1800 and 2000 °C for 2 h in He. Approximately, 200 compositions had been examined. Melting point measurement, DTA, metallography, and XRD were applied to study these systems. For DTA analysis, the alloy was placed in a graphite crucible surrounded by He at $\sim 2 \times 10^5$ Pa [2]. Among the 13 proposed diagrams (PED8868 (A)–(M)) of elements and compounds [2], 7 of them (Fig. 4.1a–g) are selected in this book. According to Ref. [3], B_4C has a homogeneity range up to the composition of $B_{13}C_2$. The authors indicated an approximate composition of $B_{\sim 4.5}C$ as the end component "B_4C" in phase diagram shown in Fig. 4.1d [1]. There is an extensive review of earlier literatures about these binary and ternary systems. Their diagrams also appeared in the final report of the series [4]. The liquidus isotherms of Fig. 4.1a are from Ref. [4]. The binary and ternary invariant points of the system are given in Table 4.1.

4.3 $TiC_{0.95}$–TiB_2

Figure 4.2 shows a quasibinary phase diagram of $TiC_{0.95}$–TiB_2 system [5]. It is one of the phase diagrams of UHTC systems published by the former Soviet Union (Kiev) Powder Metallurgy Laboratory in the 1970s [5]. The end components were consisted of chemical grade TiB_2, $TiC_{0.95}$, $TiC_{0.80}$, and $TiC_{0.68}$ that were synthesized at 1800–2200 °C in 1.3×10^{-4} Pa vacuum with raw materials Ti (>99.8%) and acetylene black. A total of 30 samples for 10 compositions spaced across the diagram were reacted at 2100–2950 °C in flowing Ar using a d.c. current and a W spiral as a supplemental external heater. The temperature was determined with a micropyrometer with a precision of ±40 °C. Sets of samples were equilibrated subsolidus at 2100 and 2300 °C. Melting temperature was monitored by short circuit of the current upon specimen melting and separation. The solubility of TiC in TiB_2 was not determined. The solubility of TiB_2 in TiC increases to ~ 7 mol% TiB_2 for the $TiC_{0.68}$–TiB_2 system. In a study of directionally solidified eutectics, Beratan [6] determined that the eutectic was at ~ 2600 °C for 65% $TiC_{0.92}$–35% TiB_2, which agrees well with the present study. Further studies can be seen in PED8868-B.

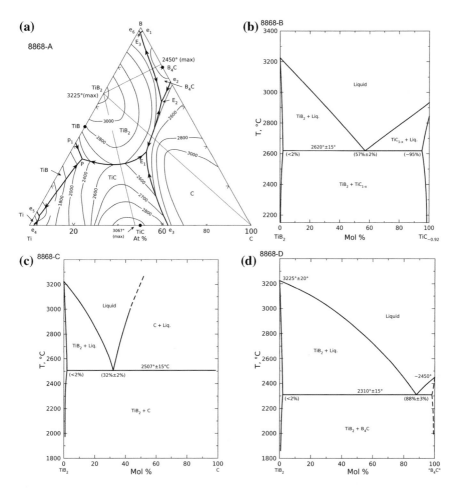

Fig. 4.1 a Liquidus projection of the B–C–Ti system. **b** Pseudobinary section of TiB$_2$–TiC$_{0.9}$. **c** Section of TiB$_2$–C. **d** Section of TiB$_2$–B$_4$C. **e** Isothermal sections of B–Ti–C system at 1700 °C. **f** Isothermal sections of B–Ti–C system at 2000 °C. **g** Isothermal sections of B–Ti–C system at 2300 °C. Reprinted with permission of The American Ceramic Society www.ceramics.org.

4.4 TiC–ZrC

Figure 4.3 shows the calculated phase equilibria of TiC–ZrC system [7]. Since the ideal and regular solution models cannot fully describe real systems, the subregular solution model [8] was used to calculate the phase equilibria using the interaction parameters. The calculation methods and the calculated diagram are also given in Ref. [8]. The calculated diagram was compared with the experimental data of Kieffer et al. [8–11]. The former was estimated to have a relative error of $\leq \pm 5$ mol%, while the latter was said to have an absolute error of ~ 10 mol%. The structure of the

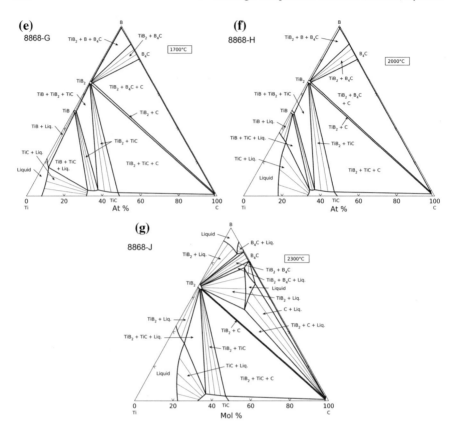

Fig. 4.1 (continued)

Table 4.1 The binary and ternary invariant points of the Ti–B–C system

Point	Temp. (°C)	Solid phases
e_1	2080	$B + B_4C$
e_2	2380	$B_4C + C$
e_3	2776	$C + TiC$
e_4	1650	$\alpha\text{-Ti} + TiC$
e_5	1540	$\beta\text{-Ti} + TiB$
e_6	2080	$TiB_2 + B$
p_1	2180	$TiB + TiB_2$
E_1	2400	$TiB_2 + C + TiC$
E_2	2240	$B_4C + TiB_2 + C$
E_3	2016	$B + B_4C + TiB_2$
P	2160	$TiB_2 + TiB + TiC$

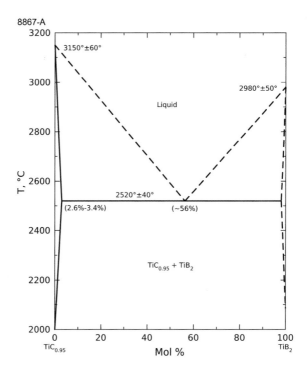

Fig. 4.2 Quasibinary phase diagram of TiC$_{0.95}$–TiB$_2$ system. Reprinted with permission of The American Ceramic Society

two monocarbides of IVB group is NaCl-type cubic, so continuous solid solution was formed at high temperatures, while miscibility gap was formed at low temperatures.

4.5 TiC–HfC

Figure 4.4a shows the calculated phase diagram of TiC–HfC system [7]. The same method as for the TiC–ZrC system was employed in calculating phase equilibria of the TiC–HfC system. Since the TiC, ZrC, and HfC are monocarbides of transition metals of the IVB group with the same NaCl-type cubic structure, they easily form solid solutions and solid solution phase relationships were established. Figure 4.4b shows the experimental phase equilibria of TiC–HfC system [12]. The phase equilibrium is from the 3-D diagram of the Ni–HfC–NbC system (see PED 9040). The melting point of HfC is 3890 °C according to the author. Figure 4.4c shows the subsolidus miscibility gap of TiC–HfC system [13]. This diagram compares the results in this work [7] with the results of the experimental work described in PED 9043 and the results in Ref. [10].

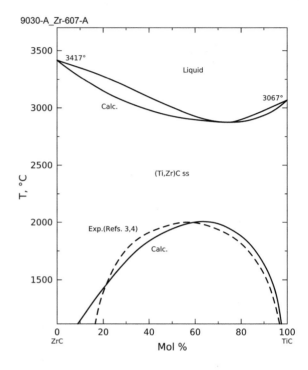

Fig. 4.3 Calculated phase equilibria of the TiC–ZrC system. Reprinted with permission of The American Ceramic Society

4.6   TiC$_{1-x}$–HfC$_{1-x}$

Figure 4.5 shows an isothermal section at 1500 °C of the Hf–Ti–C system [14]. There are two three-phase fields. One, L–(Ti, Hf) ss + H–(Ti, Hf) ss + (Hf, Ti)C$_{1-x}$ ss, is the result of the low–high transformation of the Ti–Hf ss, but is too narrow to be confirmed experimentally. The other (Hf, Ti)C$_{1-x}$ + (Ti, Hf)C$_{1-x}$ + C is the result of the partial miscibility gap in the (Ti, Hf)C$_{1-x}$ ss at high carbon concentrations. The solid solution disproportionates at ∼2050 °C on the C-rich boundary at 12–15 at.% Hf. The stability of the solid solution was further examined in PED 8989. The low-temperature form of (Hf, Ti) ss is hexagonal close-packed, while the high temperature form is body-centered cubic. The monocarbide solid solutions are NaCl-type cubic.

4.7   VC$_{0.88}$–TiC

Figure 4.6 shows the calculated phase equilibria of VC$_{0.88}$–TiC system [7]. Since the ideal and regular solution models cannot fully describe real systems, the subregular solution model [8] was used to calculate the phase equilibria using the interaction parameters. The subregular solution model takes into account the

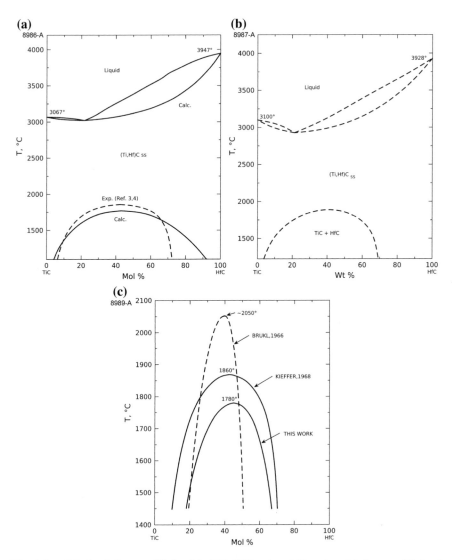

Fig. 4.4 a Calculated phase equilibria of the TiC–HfC system. **b** Experimental phase equilibria of the binary TiC–HfC system. **c** Comparison of subsolidus miscibility gap in the TiC–HfC system. Reprinted with permission of The American Ceramic Society

dependence on the temperature and composition of the interchange energy in the regular solution model. The calculation methods and the calculated diagram are also given in Ref. [9]. The calculations were estimated to have a relative error of ≤ ±5 mol%.

Fig. 4.5 Isothermal section at 1500 °C of Hf–Ti–C system. Reprinted with permission of The American Ceramic Society

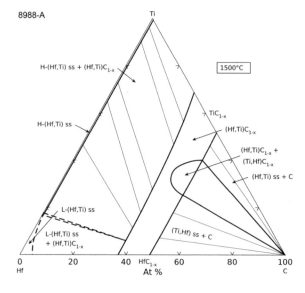

Fig. 4.6 Calculated phase equilibria of the $VC_{0.88}$–TiC system. Reprinted with permission of The American Ceramic Society

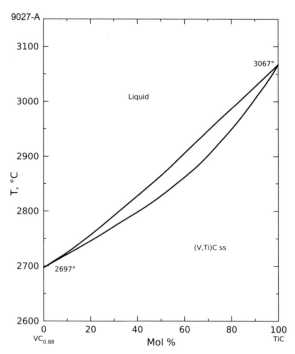

Fig. 4.7 Calculated phase equilibria of the TiC–NbC system. Reprinted with permission of The American Ceramic Society

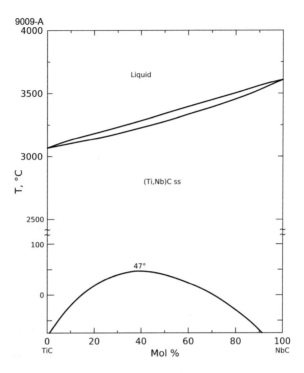

Fig. 4.8 Calculated phase equilibria of the TiC–TaC system. Reprinted with permission of The American Ceramic Society

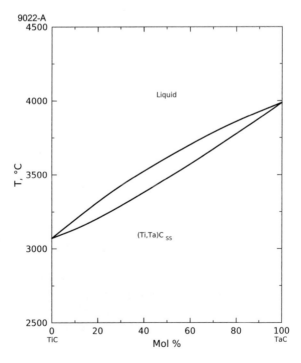

4.8 TiC–NbC

Figure 4.7 shows the calculated phase equilibria of TiC–NbC system [15]. Thermodynamic calculations applying the subregular solution model of Hardy [8] were combined with another developed thermodynamic model [7, 16] in order to calculate the exchange energies. This model takes into account the dependence on the temperature and composition of the interchange energy in the regular solution model. The calculation methods and the calculated diagram are also given in Ref. [9].

4.9 TiC–TaC

Figure 4.8 shows the calculated phase equilibria of TiC–TaC system [15]. Thermodynamic calculations applying the subregular solution model of Hardy [8] was combined with another developed thermodynamic model [7, 16] to calculate the exchange energies. This model takes into account the dependence on the temperature and composition of the interchange energy in the regular solution model. The calculation methods and the calculated diagram are also given in Ref. [9]. The critical point of the miscibility gap (not shown) was calculated to be at -23 °C.

Obviously, TiC and VC, NbC, and TaC have the same cubic structure and, respectively, form cubic continuous solid solution.

4.10 TiB_2–GdB_6

Figure 4.9 shows the pseudobinary vertical section of GdB_6–TiB_2 system [17]. It is a vertical cut from the ternary Gd–Ti–B system, where GdB_6 melts incongruently. Experimental verification of this pseudobinary system involved 28 samples of 14 compositions in 5 or 10 wt% increments across the binary. These samples were prepared from "Ch" grade GdB_6 and TiB_2 powders by mixing in ethanol in a steel mortar, cold pressing, and sintering under high-purity argon in an electric furnace at 1797 °C for 1 h. Melting point was determined by measurement of the sudden increase in direct current resistivity upon melting. The liquidus was tentative. Metallographic examination, X-ray diffraction, and microhardness measurements were used to establish the existence of the GdB_6–TiB_2 eutectic. No intermediate compound was observed, which is in agreement with the published Ti–Gd–B phase diagram [18] (isothermal section at 800 °C). The congruent melting point of GdB_4 was estimated (or derived from a previous study [19]) and the GdB_6 peritectic temperature was described as the "melting point" of GdB_6. This diagram closely resembles that of GdB_6–HfB_2 (PED 8837). This similarity is expected since both of them belong to the generalized GdB_6–$M^{IV}B_2$ group of systems. Further discussion

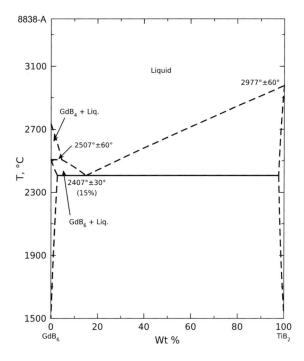

Fig. 4.9 Pseudobinary vertical section of the GdB$_6$–TiB$_2$ system. Reprinted with permission of The American Ceramic Society

of the GdB$_6$ peritectic can be found in a study of the Gd–B system by Blanks [20]. Another work [21] also confirmed that TiB$_2$ melts congruently.

4.11   TiB$_2$–TiN$_x$

Figure 4.10 shows the quasibinary T–X sections of TiN$_x$–TiB$_2$ system with different x [22]. About 13 compositions across the range were studied for each of the values of x. "Pure-grade" titanium nitride and boride powders were selected as starting materials. To prepare the nonstoichiometric TiN$_x$ materials, Ti was added to the titanium nitride powder and homogenized at 1497 °C in high-purity Ar. The specimens were prepared by grinding the appropriate powder mixtures, pressing into disks with paraffin as a binder, and sintering at 1997 °C under N$_2$ (0.1–0.2 MPa) for the TiN$_{0.93}$–TiB$_2$ system, and sintering at 1797 °C under Ar for the systems with lower values of x in the TiN$_x$. Melting points were determined by optical micropyrometer. The annealing, quenching, and melting treatments were performed in "high-grade" N$_2$ and Ar in a chamber permitting pressures up to 10 MPa.

Specimens were characterized by metallographic, X-ray, chemical analyses, and microhardness measurements. The X-ray and metallographic studies showed that the amount of TiB$_2$ that dissolves in titanium nitride increases with the degree of imperfection in the nitrogen sublattice of the TiN$_x$ phase. As the nonstoichiometry

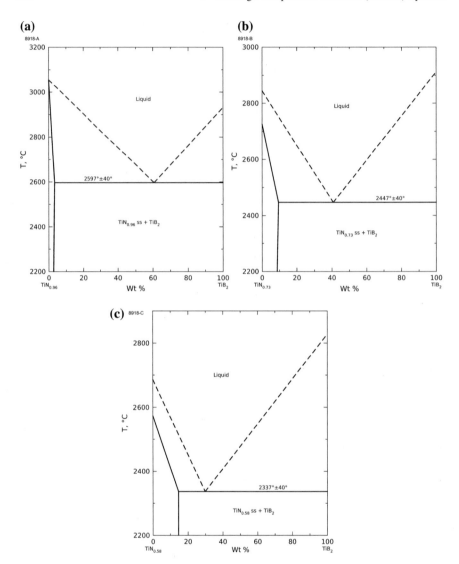

Fig. 4.10 a Quasibinary $T\text{--}X$ sections of $TiN_x\text{--}TiB_2$ system for $x = 0.96$. **b** Quasibinary $T\text{--}X$ sections of $TiN_x\text{--}TiB_2$ system for $x = 0.73$. **c** Quasibinary $T\text{--}X$ sections of $TiN_x\text{--}TiB_2$ system for $x = 0.58$. Reprinted with permission of The American Ceramic Society

of the TiN_x increases (that is, as x decreases), the eutectic temperature decreases and the composition of the eutectic shifts toward the TiN_x phase. It is noted that the liquidus curves are estimated from the eutectic and end-member melting point data. Sample collapse occurred generally up to 100 °C below the indicated liquidus temperature. Discussion, sources of the available binary data, and available crystallographic data for phases reported in the system are provided in Ref. [23].

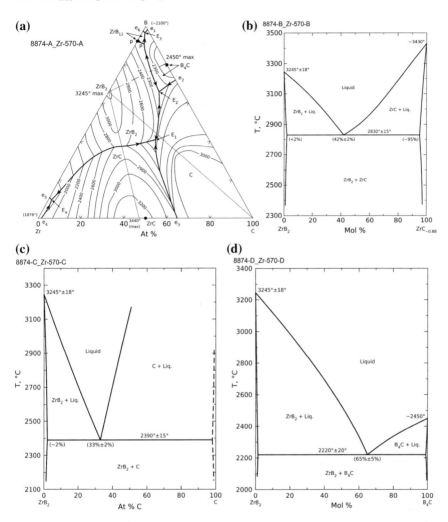

Fig. 4.11 a Liquidus projection of the B–Zr–C system. **b** Section ZrB$_2$–ZrC$_{0.88}$. **c** Section ZrB$_2$–C. **d** Section ZrB$_2$–B$_4$C. **e** Four-phase equilibrium plane of the B–Zr–C system (liquid + ZrB$_2$ ⟶ ZrB$_{12}$ + B$_4$C). Reprinted with permission of The American Ceramic Society

4.12 ZrC$_{0.88}$–ZrB$_2$ and ZrB$_2$–B$_4$C

Figure 4.11 shows the phase diagram of B–Zr–C system [1]. Figure 4.11a shows the liquidus projection; Fig. 4.11b–d shows the sections of ZrB$_2$–ZrC$_{0.88}$, ZrB$_2$–C, and ZrB$_2$–B$_4$C. Figure 4.11e shows a four-phase equilibrium plane, liquid + ZrB$_2$ → ZrB$_{12}$ + B$_4$C.

Sample and experimental preparations are the same as the system 4.2. Phase relations in the ternary system involving ZrB$_{12}$ are tentative. ZrB$_{12}$ is not observed

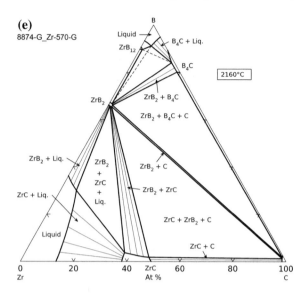

Fig. 4.11 (continued)

below ~ 1800 °C and decomposes at ~ 2160 °C. The dashed lines in Fig. 4.11e represent the phase equilibria immediately before and after phase formation/ decomposition at those temperatures. Also, the pseudobinary section of Fig. 4.11b is alternately given as PED 8875. The two diagrams agree only in the eutectic nature and the approximate eutectic composition.

There is an extensive review of earlier literature for the binary and ternary systems. These diagrams also appear in the final report of the series [4]. The liquidus isotherms in Fig. 4.11a are from the diagram published in Ref. [4]. The binary and ternary invariant points are given in Table 4.2. Twelve figures were reported, of which five are selected here.

Table 4.2 Invariant points of the binary and ternary systems

Point	Temp. (°C)	Solid phases
e_1	2080	$B_4C + B$
e_2	2375	$B_4C + C$
e_3	2911	$ZrC + C$
e_4	1833	$ZrC + \beta\text{-}Zr$
e_5	1660	$\beta\text{-}Zr + ZrB_2$
e_6	~ 2000	$B + ZrB_{12}$
p	2250	$ZrB_{12} + ZrB_2$
E_1	2360	$C + ZrC + ZrB_2$
E_2	2165	$B_4C + ZrB_2 + C$
E_3	~ 1990	$B + ZrB_{12} + B_4C$
E_4	1615	$\beta\text{--}Zr + ZrB_2 + ZrC$
P	2160	$ZrB_{12} + ZrB_2 + B_4C$

4.13 ZrC–ZrB₂

Figure 4.12 shows the quasibinary phase diagram of ZrC–ZrB$_2$ system [24]. Approximately, six evenly spaced compositions, with additional compositions at ZrB$_2$ < 15 mol%, were examined. Initial sintering was performed at 2100 °C for 4 h in vacuum (<1.33 × 10^{-8} Pa). The samples were equilibrated in flowing Ar using a d.c. current and applying a W spiral as a supplemental external heater. Subsolidus anneals were conducted at 2100 and 2400 °C. Four examples are at the eutectic temperature, while eight examples are at ~2700–3500 °C (eutectic to liquidus temperatures). Liquid formation was monitored by short circuit of the current upon specimen melting and separation. The temperature was determined using a micropyrometer to a precision of ±40 °C.

Characterization methods consisted of X-ray diffraction, chemical analysis, microhardness tests, and optical microscopy. Specimen collapse was generally observed upon formation of a few percent liquids. The quasibinary liquidus was therefore constructed from the apparent eutectic and the compound melting points. The observed eutectic at 57 mol% ZrB$_2$ and 2600 ± 40 °C can be compared with that determined by Rudy [4], which, for the ZrC$_{0.88}$–ZrB$_2$ system, showed a eutectic of 58 mol% ZrB$_2$ at 2830 ± 15 °C. Other work on directionally solidified eutectics suggested a eutectic temperature of 2500 °C.

Fig. 4.12 Quasibinary phase diagram of the ZrC–ZrB$_2$ system. Reprinted with permission of The American Ceramic Society

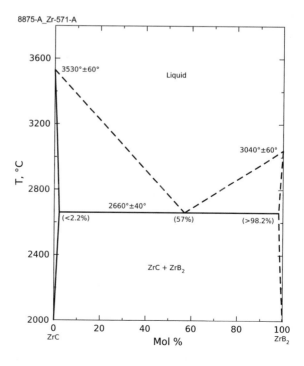

Fig. 4.13 Calculated phase equilibria of the VC$_{0.88}$–ZrC system. Reprinted with permission of The American Ceramic Society

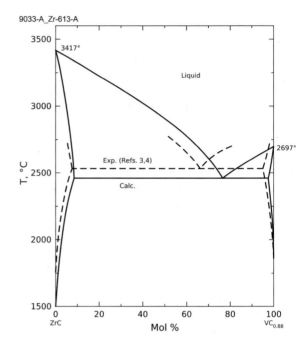

4.14 ZrC–VC$_{0.88}$

Figure 4.13 shows the calculated phase equilibria of VC$_{0.88}$–ZrC system [7]. The calculation method is also given in the system Sect. 4.9. The calculated diagram was compared to the experimental data of Refs. [10, 11]. The calculated diagram was estimated to have a relative error of $\leq \pm 5$ mol%, while the experimental data were said to have an absolute error of ~ 10 mol%.

4.15 ZrC–NbC

Figure 4.14 shows the calculated phase equilibria of ZrC–NbC system [15, 25]. Thermodynamic calculations using the subregular solution model of Hardy [8] were combined with another developed thermodynamic model [7, 16] to calculate the exchange energies. This model takes into account the dependence on the temperature and composition of the interchange energy in the regular solution model. The calculation methods and the calculated diagram were also provided in Ref. [9].

Fig. 4.14 Calculated phase equilibria of the ZrC–NbC system. Reprinted with permission of The American Ceramic Society

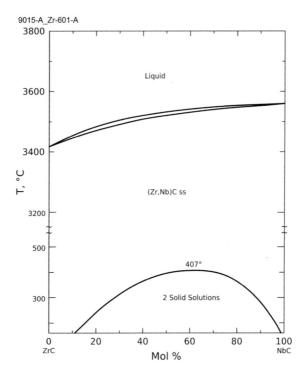

4.16   ZrC–TaC

Figure 4.15 shows the calculated phase equilibria of ZrC–TaC system [15, 25]. The calculation method is also given in system 4.15. The critical point of the miscibility gap was not provided.

4.17   ZrB$_2$–SiC

Figure 4.16 shows the pseudobinary phase diagram of ZrB$_2$–SiC system [26]. Using reagent-grade SiC and ZrB$_2$, nine evenly spaced compositions were equilibrated at ∼2000–2700 °C in flowing Ar using a direct current and a W spiral as a supplemental external heater. Samples were equilibrated at 2000 °C and also heated to melting to determine the liquidus minimum. Liquid formation was monitored by short circuit of the current upon specimen melting. This system is pseudobinary owing to the peritectic decomposition of SiC at 2760 °C [27]. The system still shows simple eutectic behavior, and the mutual solid solubility is very small. The liquidus was extrapolated by the authors from the apparent experimental eutectic to the known melting or decomposition temperatures. The melting point of ZrB$_2$ has been reported 3245 °C [28], but it is shown 3050 °C in this phase diagram [26].

Fig. 4.15 Calculated phase
equilibria of the ZrC–TaC
system. Reprinted with
permission of The American
Ceramic Society

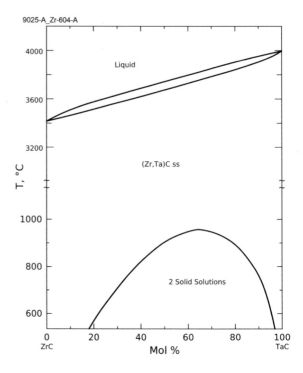

Fig. 4.16 Pseudobinary
phase diagram of ZrB$_2$–SiC
system. Reprinted with
permission of The American
Ceramic Society

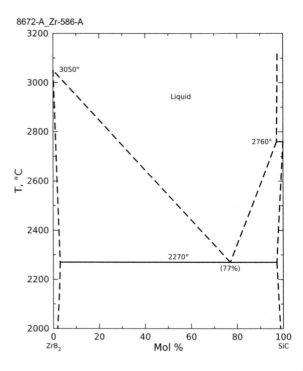

Fig. 4.17 Binary phase equilibria of the ZrN$_{0.96}$–ZrB$_2$ system under 1 MPa N$_2$. Reprinted with permission of The American Ceramic Society

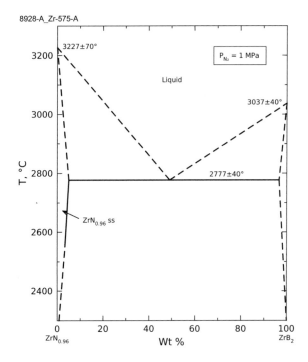

4.18 ZrB$_2$–ZrN$_{0.96}$

Figure 4.17 shows the binary phase equilibria of ZrN$_{0.96}$–ZrB$_2$ system under 1 MPa N$_2$ [29]. The raw materials were zirconium nitride and zirconium diboride powders. The disks were sintered at 1997 °C under an excess ultrapure nitrogen pressure of $(0.10–0.2) \times 10^5$ Pa. Metallurgical analysis showed the presence of a two-phase alloy when more than 5 wt% ZrB$_2$ was present. However, below 5 wt% ZrB$_2$, an increase in unit cell parameters of the nitride phase and its microhardness in ZrB$_2$ alloys indicated the formation of a ZrN$_{1-x}$B$_{2x}$-type solid solution. The liquidus is marked as dashed line to indicate uncertainty due to the effects of incongruent evaporation of the components from the melt. There was dispute with the melting point of ZrN, which depends on the nitrogen pressure [30].

4.19 ZrB$_2$–W$_2$B$_5$

Figure 4.18 shows the experimental phase equilibria of ZrB$_2$–W$_2$B$_5$ system [31] (selected from the Zr–W–B element system). (a) is the isothermal section at 1400 °C; and (b) is vertical section ZrB$_2$–W$_2$B$_5$. About 73 compositions were prepared for solid-state studies and 48 compositions for melting point studies [2]. The starting

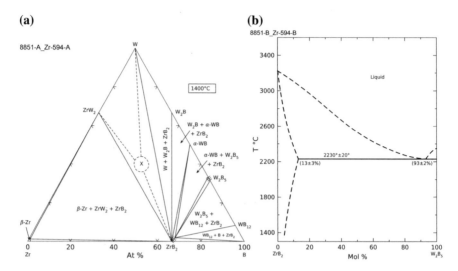

Fig. 4.18 The experimental phase equilibria of ZrB_2–W_2B_5 system: **a** Is the isothermal section of Zr-W-B element system at 1400 °C. **b** Is vertical section ZrB_2–W_2B_5. Reprinted with permission of The American Ceramic Society

materials were the following powders: ZrH_2, high-purity W, and 99.55% B. ZrB2 was prepared by direct reaction of the element powders (contained 0.16% C).

The equilibria in the region ZrB_2–ZrW_2–W–W_2B, see Fig. 4.18a (isothermal section at 1400 °C), are not firmly established because of the sluggish reaction rate of formation of the ternary phase X. Single-phase specimens of X could not be prepared.

The ZrB_2 was in equilibrium with all of the tungsten borides, and binary systems were formed. Figure 4.18b shows the vertical section of ZrB_2–W_2B_5. For ZrB_2–WB, the eutectic is at 2530 °C and $\sim$70 mol% WB, for ZrB_2–W_2B the eutectic is at 2480 °C and $\sim$60 mol% W_2B, and for ZrB_2–W the eutectic is at 2250 °C and $\sim$21 mol% W. The diagrams also appear in the final report of the series [4].

Romans and Krug [32] reported the highest boride of tungsten to be WB_4, but they said it may contain more B, up to WB_5. They found it to be hexagonal and published an indexed X-ray pattern. They did not confirm the WB_{12} phase. Post and Glaser [33] reported a ZrB_{12} phase and gave a detailed crystal structure report but did not report preparation conditions.

4.20 Hf–Zr–C (Hf, Zr)C_{1-x} ss)

Figure 4.19 is taken from the Hf–Zr–C system [14]. Figure 4.19a shows the liquidus projection and the dashed line indicates the maximum solidus temperatures of the (Zr, Hf)C solid solution; II_a is the four-phase reaction plane, β-

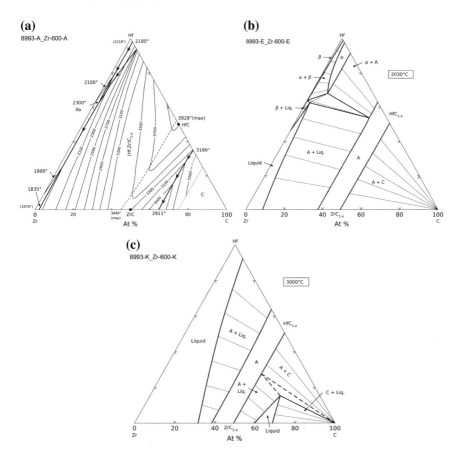

Fig. 4.19 **a** Liquidus projection of Hf–Zr–C system. **b** Isothermal sections of Hf–Zr–C system at 2030 °C. **c** Isothermal sections of Hf–Zr–C system at 3000 °C. Reprinted with permission of The American Ceramic Society

(Hf, Zr) + (Hf, Zr)C$_{1-x}$ ⟷ Liquid + α-Hf(C, Zr). Figure 4.19b, c shows the isothermal sections at 2030 °C (four-phase reaction plane) and 3000 °C, respectively. There are 15 phase figures reported [14], of which three diagrams are introduced here.

Most samples were prepared in the manner described in PED 9031. The A is cubic (Hf, Zr)C$_{1-x}$ ss which is a continuous solid solution below 3300 °C. α is low-temperature hexagonal close-packed structure of Zr and Hf and their solid solutions; and β is high temperature body-centered cubic structure of Zr and Hf and their solid solutions.

Fig. 4.20 Calculated phase equilibria of the ZrC–HfC system. Reprinted with permission of The American Ceramic Society

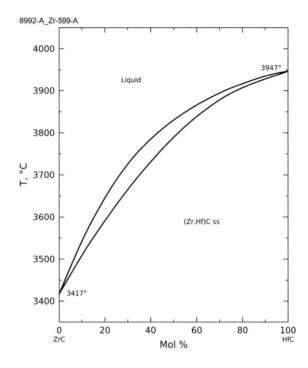

8992-A_Zr-599-A

4.21 HfC–ZrC

Figure 4.20 shows the calculated phase equilibria of ZrC–HfC system [7]. Since the ideal and regular solution models cannot fully describe real systems, the subregular solution model [8] was used to calculate the phase equilibria using the interaction parameters. The subregular solution model takes into account the dependence on the temperature and composition of the interchange energy in the regular solution model. The calculation methods and the calculated diagram are also provided in Ref. [9]. The calculations were estimated to have a relative error of $\leq \pm 5$ mol%.

4.22 HfC–TaC

Figure 4.21 shows the calculated phase equilibria of HfC–TaC system [15, 25]. The calculation methods are also given in system 4.15. The calculation methods and the calculated diagram are also provided in Ref. [9].

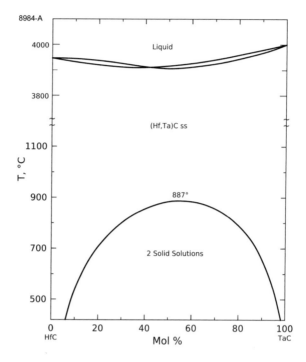

Fig. 4.21 Calculated phase equilibria of the HfC–TaC system. Reprinted with permission of The American Ceramic Society

4.23 HfB$_2$–HfC$_{0.9}$ and HfB$_2$–B$_4$C

Figure 4.22 is taken from the Hf–B–C system [1]. There are 14 figures proposed by the author, 7 of which were included in here. Figure 4.22a shows the liquidus projection, and Fig. 4.22b–d shows the sections of HfB$_2$–HfC$_{0.9}$, HfB$_2$–C, and HfB$_2$–B$_4$C, respectively. Figure 4.22e–g shows the isothermal sections at 1800, 2000, and 2400 °C, respectively.

The specimens were prepared from Hf powder, Hf sponge, spectrographic graphite, and B powder. The HfB$_2$ was prepared from the elements in a two-step process to avoid the violent reaction caused by direct reaction between the elements. First, a master alloy containing 85 at.% B was prepared, then mixed intimately with the appropriate amount of Hf, and reacted under He for 2 h at temperatures between 1800 and 2000 °C. Most of the samples were prepared by hot pressing of well-blended mixtures in graphite dies followed by surface grinding, while some samples were made by electron beam melting or arc melting.

Melting point study, DTA, metallography, and X-ray diffraction were applied. Phase relations in the high-Hf corner are complex due to the differing solubilities of B and C in the Hf ss phases. B$_4$C has a homogeneity range up to the composition B$_{13}$C$_2$ according to Ref. [3]. The binary and ternary invariant points are given in Table 4.3. The two four-phase equilibrium planes are given in Table 4.4.

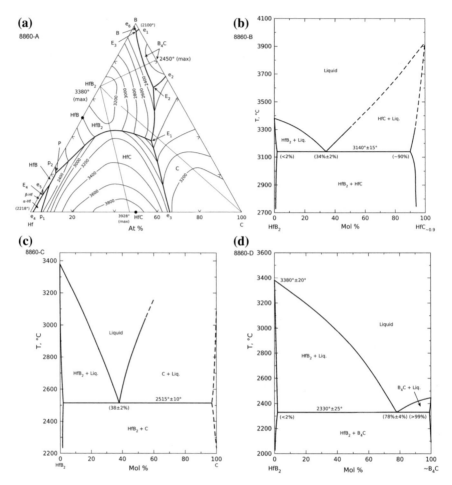

Fig. 4.22 a Liquidus projection of Hf–B–C system. **b** Section of HfB$_2$–HfC$_{0.9}$. **c** Section of HfB$_2$–C. **d** Section of HfB$_2$–B$_4$C. **e** Isothermal sections of Hf–B–C system at 1800 °C. **f** Isothermal sections of Hf–B–C system at 2000 °C. **g** Isothermal sections of Hf–B–C system at 2400 °C. Reprinted with permission of The American Ceramic Society

4.24 HfB$_2$–GdB$_6$

Figure 4.23 shows the pseudobinary vertical section of GdB$_6$–HfB$_2$ system [34]. In determining this pseudobinary system (a vertical cut from the Gd–Hf–B ternary system, where GdB$_6$ melts incongruently), 26 experimental samples of 13 compositions in 5 or 10 wt% increments across the binary were used. The samples were sintered under high-purity argon in an electric furnace at 1797 °C for 1 h. Melting point was determined by measurement of the sudden increase in direct current

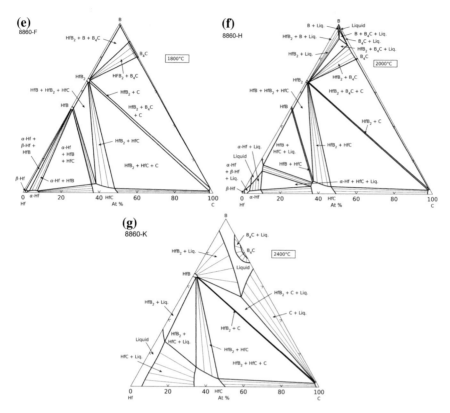

Fig. 4.22 (continued)

Table 4.3 The binary and ternary invariant points

Point phase	Temp. (°C)	Solid phases
e_1	2080	B + B$_4$C
e_2	2380	B$_4$C + C
e_3	3180	C + HfC
e_4	2180	α-Hf + β-Hf
e_5	1880	β-Hf + HfB
e_6	2065	HfB$_2$ + B
p_1	2360	α-Hf + HfC
p_2	2100	HfB + HfB$_2$
E_1	2480	C + HfC + HfB$_2$
E_2	2260	B$_4$C + C + HfB$_2$
E_3	1950	B + B$_4$C + HfB$_2$
E_4	1850	β-Hf + HfB + α-Hf
P	2050	HfB + HfB$_2$ + HfC

Table 4.4 The two
four-phase equilibrium planes

Temp. (°C)	Phases in equilibrium
1850	Liquid $\longleftrightarrow$ α-Hf + β-Hf + HfB
1940	Liquid + HfC $\longleftrightarrow$ HfB + α-Hf
1950	Liquid $\longleftrightarrow$ HfB$_2$ + B$_4$C + B
2050	Liquid + HfB$_2$ $\longleftrightarrow$ HfB + HfC
2260	Liquid $\longleftrightarrow$ HfB$_2$ + B$_4$C + C
2480	Liquid $\longleftrightarrow$ HfB$_2$ + HfC + C

Fig. 4.23 Pseudobinary
vertical section of GdB$_6$–
HfB$_2$ system. Reprinted with
permission of The American
Ceramic Society

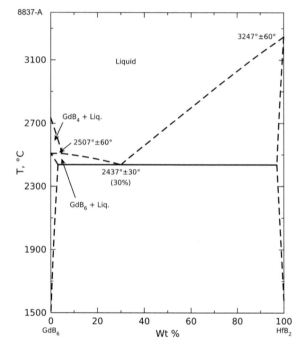

resistivity upon melting. The atmosphere used in the melting point studies was not
described, which lead to a tentative liquidus curve.

Metallographic examination, X-ray diffraction, and microhardness measurements were used to establish the existence of the GdB$_6$–HfB$_2$ eutectic. No intermediate compound was observed, which is in agreement with the published
Gd–Hf–B phase diagram [18] (isothermal section at 800 °C). The solubility limits
of the end members were not determined but were below the 5 wt% level (limit of
the experimental compositions used). The congruent melting point of GdB$_4$ was
estimated (or derived from a previous study [19]), and the GdB$_6$ peritectic temperature was described as the "melting point" of GdB$_6$. This diagram closely
resembles that of GdB$_6$–TiB$_2$ (PED 8838). This similarity is expected since both
phases belong to the generalized GdB$_6$–MB$_2$ group of systems. Further discussion
of the GdB$_6$ peritectic can be found in a study of the Gd–B system by Blanks [20].
Another work [21] confirmed that HfB$_2$ melts congruently.

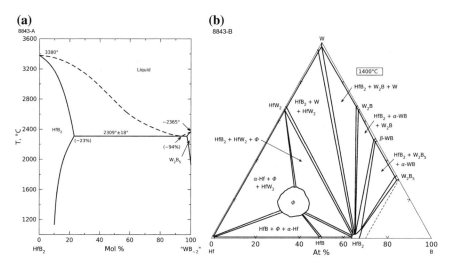

Fig. 24.4 a Vertical section of HfB$_2$–W$_2$B$_5$ system. **b** Isothermal section of HfB$_2$–W$_2$B$_5$ system at 1400 °C (φ = ternary compound of unknown composition). Reprinted with permission of The American Ceramic Society

4.25 HfB$_2$–W$_2$B$_5$

Figure 4.24 shows (a) the vertical section of HfB$_2$–W$_2$B$_5$ system and (b) the isothermal section at 1400 °C of HfB$_2$–W$_2$B$_5$ system [35]. HfB$_2$ was prepared by mixing intimately with the appropriate amount of Hf, cold compacting, and heating in a Ta container under He for 2 h at temperatures between 1800 and 2000 °C. The W$_2$B$_5$ was prepared by cold compact elements, first heating to 1200 °C in a Ta container in a carbon pot furnace, and completing the reaction by a 2-hour vacuum treatment at 1750 °C. The W$_2$B$_5$ contained significant C (0.12%).

The ternary alloys were prepared by hot pressing the starting material mixtures in the graphite dies. Carbon was removed from the surface by grinding before the heat treatment at 1400 °C for 100 h under vacuum (<6.7 × 10^3 Pa).

Some alloys were prepared by arc melting under He in a water-cooled Cu hearth [2].

Some of the products of the heat treatments were studied by melting point investigation [2], especially for the section shown in Fig. 4.24a. The melting points were determined under 2.3 × 10^5 Pa He. Figure 4.24 shows (a) the vertical section and (b) the isothermal section at 1400 °C of HfB$_2$–W$_2$B$_5$ system [35].

A ternary compound (designated φ) has been reported to occur but the composition and structure are not investigated here [35]. See PED 8841 where the same sort of compound also occurs.

There is an extensive review of earlier literature for the binary and ternary systems. The diagrams also appear in the final report of the series [4].

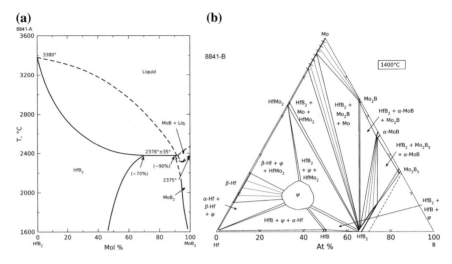

Fig.4.25 a Pseudobinary section of HfB$_2$–MoB$_2$ of the Hf–Mo–B system. **b** Isothermal section of Hf–Mo–B system at 1400 °C. Reprinted with permission of The American Ceramic Society

4.26   HfB$_2$–MoB$_2$

Figure 4.25 is taken from the Hf–Mo–B system; Fig. 4.25a shows the pseudobinary section of HfB$_2$–MoB$_2$, and Fig. 4.25b shows the isothermal section at 1400 °C [35].

Compositions along the pseudobinary HfB$_2$–MoB$_2$ were heated to 2000 °C for 25 h under high-purity He after the 1400 °C treatment. Some alloys were prepared by arc melting under He on a water-cooled Cu hearth.

Some of the heat-treated products were studied by melting point investigation [2], especially for the section shown in Fig. 4.25a. The melting points were determined under 2.3 × 10^5 Pa He. A ternary compound (designated φ) is reported to occur but the composition and structure were not investigated here [35]. Rogl et al. [36] report a phase, Hf$_9$Mo$_4$B, as hexagonal with a = 0.858 nm and c = 0.849 nm. The diagrams also appear in the final report of the series [4]. Figure 4.24b indicates that the HfB$_2$ is in equilibrium state with all of the molybdenum borides.

4.27   TiN–AlN

Figure 4.26 shows the diagram of Ti–Al–N system at 1000 °C [37]. Diffusion pairs of both types were obtained by depositing titanium on an aluminum nitride plate in an electric arc furnace as well as by diffusion welding of AlN and Ti plates in a vacuum using a DSVU apparatus followed by prolonged vacuum annealing (up to

12596-A

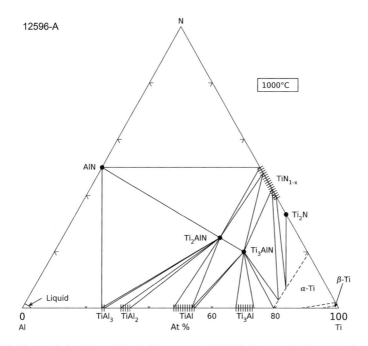

Fig. 4.26 Phase relationships of Ti–Al–N system at 1000 °C. Reprinted with permission of The American Ceramic Society

200 h) at 1000 °C. The phase equilibria of the Ti–Al–N elements system were obtained through complex experiment [38] and calculation of Gibbs energies. Except Ti–Al alloy, two ternary compounds were formed, Ti_2AlN and Ti_3AlN. The system is composed of the ternary compounds. The system has reached equilibrium since the variation of Gibbs energies is not obvious [39]. Two compounds are in equilibrium state with TiN_{1-x}. The line between AlN and TiN is a binary tie line.

4.28 ZrN–AlN

Figure 4.27 shows the subsolidus phase diagram of Al–Zr–N system at 1000 °C, which was obtained by experiment and calculation of Gibbs free energies [40]. Detailed relations for the ZrN solid solution phase were not determined. There are two ternary compounds, Zr_3AlN and $Zr_5Al_3N_{1-x}$ [38]. The latter was in equilibrium state with Zr–Al alloy. Gibbs formation energies were determined at 1000 °C for the ternary compounds, as follows: $1/6\Delta fG(Zr_3AlN) = -76.0$ kJ/mol and $[1/(9 - x)]\Delta fG(Zr_{54}Al_3N_{1-x}) = -63.0$ kJ/mol. $Zr_5Al_3N_{1-x}$ forms a set of phase fields together with Zr_2Al_3 and β-Zr ss close to Al–Zr boundary.

Phase relations for the 1000 °C section do not agree with those presented in Ref. [38] at the same temperature (where Zr_3AlN–AlN and Zr_2Al_3–AlN tie lines are

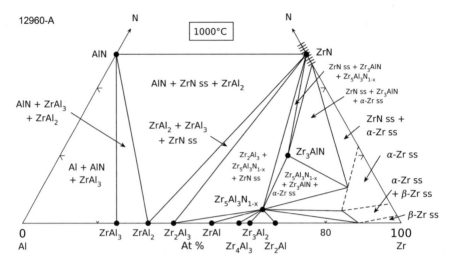

Fig. 4.27 Subsolidus phase diagram Al–Zr–N system at 1000 °C. Reprinted with permission of The American Ceramic Society

stable) but are very similar to the isothermal section at 1300 °C reported in the same work [41]. The phase diagram includes a three-phase coexistent triangle of AlN–ZrN ss–ZrAl$_2$ and is applicable to developing mechanically strong high temperature composite materials [40].

4.29   ZrN–ZrO$_2$–Y$_2$O$_3$

Figure 4.28 shows the tentative phase diagram of ZrN–ZrO$_2$–Y$_2$O$_3$ system at 1750 °C. Five compositions were prepared to investigate the phase relation of the ZrN–ZrO$_2$–Y$_2$O$_3$ system. The mixture of the compositions was hot pressed at 1750 °C in nitrogen under 16–31 MPa for 1 h [42]. The phases were analyzed by XRD. The results show that the phase relationship of ZrN based system is similar to that of the SiC-based system [43]. Like SiC, the ZrN did not react with end-point oxides. The ZrN coexists with and is in equilibrium state with the yttrium–zirconium oxides solid solution. Then, the tentative phase diagram of the ZrN–ZrO$_2$–Y$_2$O$_3$ system is proposed.

4.30   NbC–VC–Ni

Figure 4.29 shows the ternary phase diagram of Ni–NbC–VC system [44].

An unstated number of samples were prepared from 99.8% Ni, 99.98% Nb, 99.94% V, and spectrum-pure C (graphite). The alloys were arc melted in a

Fig. 4.28 Tentative phase diagram of the ZrN–ZrO₂– Y₂O₃ system

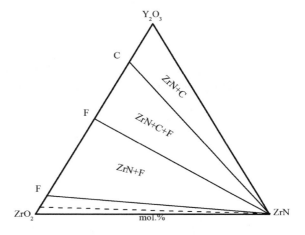

Fig. 4.29 Ternary phase diagram of the Ni–NbC–VC system. Reprinted with permission of The American Ceramic Society

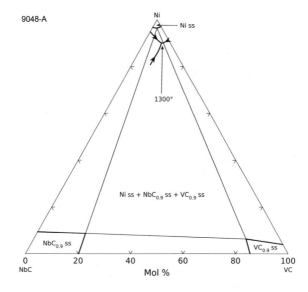

water-cooled copper hearth under an argon atmosphere. Samples were also annealed in vacuum at 1110 °C for 10 h. The diagram was determined using DTA in He (heating rate 80°), metallographic microscopy, and X-ray diffraction (CuKα) techniques. The annealing in vacuum did not lead to a change in the melting temperatures. The ternary eutectic is at 1300 °C, 91% Ni, 3% $NbC_{0.9}$, and 6% $VC_{0.9}$. A schematic three-dimensional liquidus diagram is given in the original work but the copy was not sufficiently clear to enable reproduction here.

4.31 NbC–VC

Figure 4.30 shows the phase diagram of VC–NbC system [44]. This study was performed by the same methods as in Sect. 4.30. The solid solution was formed since the same cubic structure of VC and NbC with monocarbide of M^VC. It is noted that the phase rule requires a narrow two-phase "Liq + (V, Nb)C ss" region, not shown by the authors [44] and presumably narrower than the precision (not stated but often as much as ± 60 °C in this temperature range).

4.32 NbB₂–B₄C

Figure 4.31 shows the quasibinary phase diagram of B_4C–NbB_2 system [45]. About 12 compositions were prepared. The preparation and analysis methods are the same as stated in Sect. 4.34. Very little, if any, solid solution was detected in the work, but the authors indicated that there was a small amount on both sides of the diagram.

Specimens containing NbB_2 + liquid collapsed on the formation of liquid, and thus the authors' liquidus is a smooth curve from the experimental eutectic to the known melting point of NbB_2, $\sim 3000 \pm 40$ °C.

Fig. 4.30 Phase diagram of the VC–NbC system. Reprinted with permission of The American Ceramic Society

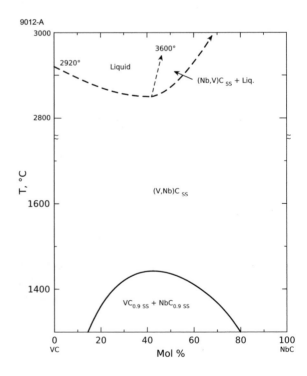

Fig. 4.31 Quasibinary phase diagram of B$_4$C–NbB$_2$ system. Reprinted with permission of The American Ceramic Society

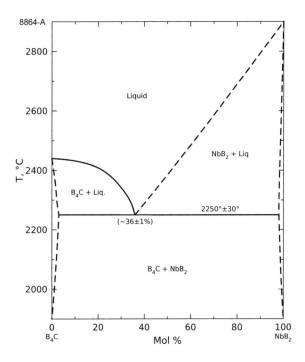

4.33   TaB$_2$–B$_4$C

Figure 4.32 shows the quasibinary phase diagram of B$_4$C–TaB$_2$ system [45]. The preparation and analysis methods are the same as in Sect. 4.32. Specimens containing TaB$_2$ + liquid collapsed above the eutectic temperature, and thus the authors' liquidus is a smooth curve from the experimental eutectic to the known melting point of TaB$_2$, $\sim 3100° \pm 60$ °C.

4.34   VB$_2$–B$_4$C

Figure 4.33 shows the quasibinary phase diagram of VB$_2$–B$_4$C system [45].

About 12 compositions were prepared from B$_4$C, which had been purified to ~ 0.2 wt% of free C by vacuum annealing at 2000 °C, and VB$_2$. Samples with high-B$_4$C content were heated indirectly, while those samples consisting mainly of VB$_2$ were melted by direct passage of current. The samples were studied by X-ray diffraction and metallographic analysis. Very little, if any, solid solution was detected in the work but the authors indicated that there was a small amount on both sides of the diagram. The liquidus temperatures at >70 mol% VB$_2$ could not be determined experimentally due to specimen collapse. The authors' liquidus is a

Fig.4.32 Quasibinary phase diagram of B$_4$C–TaB$_2$ system. Reprinted with permission of The American Ceramic Society

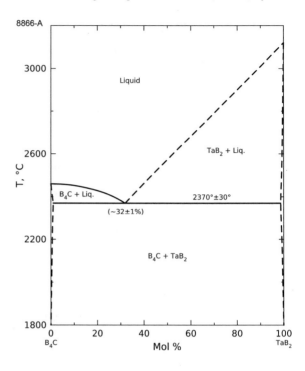

Fig. 4.33 Quasibinary phase diagram of B$_4$C–VB$_2$ system. Reprinted with permission of The American Ceramic Society

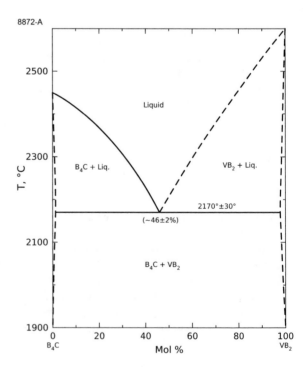

smooth extrapolation from the experimental eutectic to the known melting point of VB$_2$, $\approx$2600° $\pm$ 60 $^\circ$C.

4.35 VB$_2$–VC$_{0.88}$

The starting components were VC$_{0.88}$, which was synthesized from the elements, and reagent-grade VB$_2$. The powders were mixed using a steel mortar and pestle with ethanol as a medium. Initial sintering was performed at 1800 $^\circ$C for 2 h in vacuum ($\approx$10^{-2} Pa). 1 wt% Ni was added in order to aid sintering, which evaporated completely during heat treatment. Spark cutting was used to produce samples for higher temperature studies. These samples were equilibrated in flowing Ar using a d.c. current and a W spiral as a supplemental external heater. Subsolidus equilibration studies involved nine different compositions heated at 1800 and 2000 $^\circ$C. Liquid formation was monitored by short circuit of the current upon specimen melting and separation. The temperature was determined with a micropyrometer to a precision of $\pm$30 $^\circ$C. The pyrometer was sighted in a hole simulating blackbody conditions.

Characterization methods consisted of X-ray diffraction, chemical analysis, microhardness tests, and optical microscopy.

Specimen collapse was observed on the formation of a few percent liquids, and thus the authors' liquidus is tentative. It is drawn from the experimental eutectic point to the melting points of two end compound members.

Figure 4.34 shows the quasibinary phase diagram of VC$_{0.88}$–VB$_2$ system [46]. The solubility of VB$_2$ in VC$_{0.88}$ was 4.3 mol% at 2000 $^\circ$C and almost undetectable at 1800 $^\circ$C. The solubility of VC$_{0.88}$ in VB$_2$ was negligible.

4.36 W$_2$B$_5$–B$_4$C, W$_2$B–W$_2$C, WB–W$_2$C, etc.

The specimens were prepared from the following powders: element W, B powders, and spectrographic graphite. Both the WC and W$_2$B$_5$ were prepared by cold compact the element powders, sintering in a graphite element furnace and purifying. Approximately, 40 ternary samples were prepared with additional compositions along the examined pseudobinary sections. All of the samples were prepared by hot pressing in graphite dies at 1800–2200 $^\circ$C.

The melting point tests were performed under He at atmospheric pressure. The DTA was also performed under He using graphite containers, and the TaC is served as a standard [2]. For solid-state studies, the specimens were heat treated at 1500 $^\circ$C for 140 h and at 2000 $^\circ$C for 6 h. For metallographic studies, each composition was also arc melted under He. X-ray diffraction using Cu Kα radiation

Fig. 4.34 Quasibinary phase diagram of VC$_{0.88}$–VB$_2$ system. Reprinted with permission of The American Ceramic Society

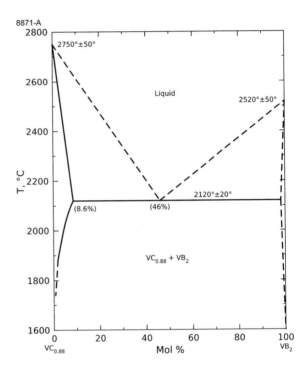

Table 4.5 Ternary eutectics of B–W–C system

Point	Temp. (°C)	Solid phases
E_1	1950	WB$_{\sim 4}$ + B + B$_4$
E_2	2180	W$_2$B$_5$ + C + B$_4$C
E_3	2240	WB + W$_2$B$_5$ + C
E_4	2300	W$_2$C + WC + WB
E_5	2305	W$_2$C + W$_2$B + WB
E_6	2355	W + W$_2$C + W$_2$B

was applied to analyze each of the specimens. Six ternary eutectics appear in the system as given in Table 4.5.

The binary invariant points are listed in Table 4.6.

Fourteen phase diagrams including PED8873 (A) to (N) were proposed by the author [47]. In addition to the three four-phase equilibria shown in PED8873 (C), (D), and (F), there is a reaction of Liquid + WC$_{1-x}$ → W$_2$C + WC equilibrium at 2570 °C. There are five reactions associated with pseudobinary equilibria as shown in PED8873 (J)–(N). The diagrams also appear in the final report of the series [4].

An earlier study [48] was performed on 45 compositions that were sintered at 1300–1700 °C and homogenized under vacuum. That study [48] presents an

Table 4.6 Binary invariant points

Point	Temp. (°C)	Solid phases
e_1	2080	B + B_4C
e_2	1970	WB_4 + B
e_3	2220	W_2B_5 + B_4C
e_4	2275	W_2B_5 + C
e_5	2340	WB + W_2B_5
e_6	2580	WB + W_2B
e_7	2360	WB + C
e_8	2600	W + W_2B
e_9	2330	WB + W_2C
e_{10}	2370	W_2B + W_2C
e_{11}	2380	B4C + C
e_{12}	2710	W + W_2C
e_{13}	2720	WC_{1-x} + WC
e_{14}	2735	W_2C + WC_{1-x}
p_1*	2070	W_2B_5 + WB_4
p_2	2776	C + WC

Note The temperature value for p_1 is 2070 °C given in this table, but 2020 °C is given in PED 8823 by the same author

isothermal section at 1700 °C, which is in agreement with the series shown here [47].

Figure 4.35 shows the 8 diagrams taken from the 14 phase diagrams of B–W–C system [47]. Figure 4.35a shows the liquidus projection; II labels the temperatures of class II four-phase equilibria; dotted lines correspond to the pseudobinary sections shown in Figs. 4.35d–h. Figure 4.35b shows the isothermal sections at 1500 °C; Fig. 4.35c shows the plane of four-phase equilibrium, Liquid + W_2B_5 $\longleftrightarrow$ WB$_{\sim 4}$ + B_4C at 2000 °C: Fig. 4.35d–h shows the pseudobinary sections of W_2B_5–B_4C, W_2B_5–C, WB–C, W_2B–W_2C, and WB–W_2C at 2000 °C, respectively.

4.37 TiC–HfC–VC

Figure 4.36 shows the isothermal section of TiC–HfC–VC system at 2050 °C [49]. Point "P" is the calculated critical point.

The diagram is based on observations of ~ 30 ternary and 25 binary compositions. Starting materials were metal carbide powders prepared via carburization of either metal oxide or metal powders of unspecified purity at about 2000 °C. All compositions contained an added 1% Co to aid in solid-state diffusion rates. Mixtures were hot pressed and then annealed under hydrogen at temperatures for an additional 28 h to homogenize the solid solutions. Phase identification and lattice

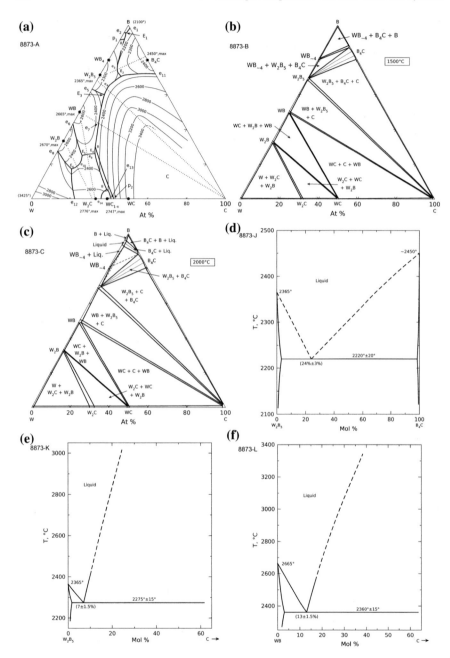

Fig. 4.35 **a** Liquidus projection of the B–W–C system. **b** Isothermal sections at 1500 °C. **c** Plane of four-phase equilibrium, liquidus + $W_2B_5 \longrightarrow WB_{\sim 4} + B_4C$ at 2000 °C. **d** Pseudobinary sections of W_2B_5–B_4C. **e** Pseudobinary sections of W_2B_5–C. **f** Pseudobinary sections of WB–C. **g** Pseudobinary sections of W_2B–W_2C. **h** Pseudobinary sections of WB–W_2C. Reprinted with permission of The American Ceramic Society

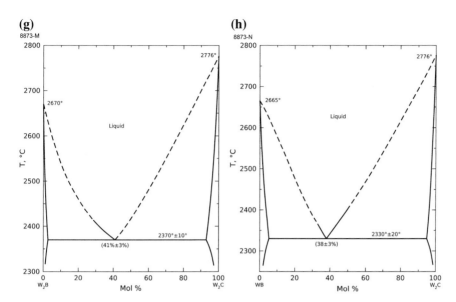

Fig. 4.35 (continued)

parameter determination were performed by X-ray diffraction. Lattice parameters of all solid solutions are presented graphically.

The critical point given in the figure was obtained through analysis of the solid solution data in terms of an ideal solution model. Since reproduction of the experimentally determined tie lines with this model was fair, only the experimental data are reproduced here.

4.38 TiC–HfC–WC

Figure 4.37 shows the experimental phase diagram of TiC–HfC–WC system [13]. Figure 4.37a shows the liquidus projection. Figure 4.37b shows the solidus phase boundaries at different temperatures. Figure 4.37c–f shows the experimental isothermal sections at (c) 1540 °C, (d) 2990 °C, (e) 3030 °C, and (f) 3100 °C, respectively. Figure 4.37g shows the isothermal section at 1500 °C comparing observed and calculated results. Figure 4.37h shows the calculated isothermal section at 1600 °C and the calculated solubilities of WC in (Ti, Hf)C at temperatures from 1600 to 2300 °C [13]. δ = (Hf, Ti, W)C ss, δ_1 and δ_2 = (Ti, Hf)C ss, and ε = hexagonal WC. P_c is the location of the critical point at $(TiC)_{0.27}(HfC)_{0.41}(WC)_{0.32}$ where the $\varepsilon + \delta_1 + \delta_2 + C$ field degenerates into one limiting tie line originating at P_c.

Starting materials were commercial powders of TiC, HfC, and WC. Mixtures of these materials were heat treated at 1500 °C for 12 h under Ar and He for a rapid

Fig. 4.36 Isothermal section
TiC–HfC–VC system at
2050 °C. Reprinted with
permission of The American
Ceramic Society

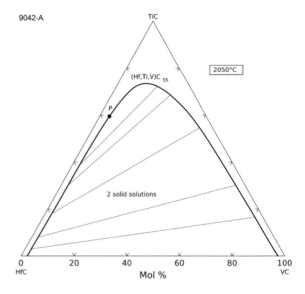

quench. Equilibration treatments at 1500 °C were under vacuum (1.3×10^{-3} Pa)
for 120 h, then at higher temperatures (1750 °C, 48 h and 2000 °C, 8 h) under Ar.

The system was investigated using melting point studies, differential thermal
analysis, X-ray diffraction, and metallographic techniques from 1550 °C to the
melting temperature [2]. Lattice parameters are given for the various phases.

A pseudobinary miscibility gap forms within the TiC–HfC system below 1780 °C
(see PED. 8989). Extension into the pseudoternary system with WC causes the gap to
move to higher temperatures, reaching a critical point at 1800 °C and
$(TiC)_{0.55}(HfC)_{0.45}(WC)_{0.05}$ (see Fig. 4.37b). The phase boundaries formed by interaction between the cubic monocarbide solid solution and the ternary miscibility gap
were studied at 1540 °C using $(TiC)_{0.27}(HfC)_{0.41}(WC)_{0.32}$. Alloys in this region
decompose upon cooling to give two cubic phases and hexagonal WC.

Some aspects of the phase relationships were calculated using a regular solution
model and assumed interaction energies, see Fig. 4.37g, h.

4.39   TiC–HfC–(MoC)

Figure 4.38 shows the phase diagram of TiC–HfC–(MoC) system [50], where
(MoC) represents a hypothetical compound at 50 at.%. Figure 4.38a shows the
liquidus projection; the dashed line represents the bottom line of a flat melting
trough. Figure 4.38b, c shows the comparison between experimental and calculated
isothermal data for the ternary miscibility gap at 1500 and 1650 °C. The P_c is the
critical point on the binodal where the miscibility gap closes and separates from the
solvus.

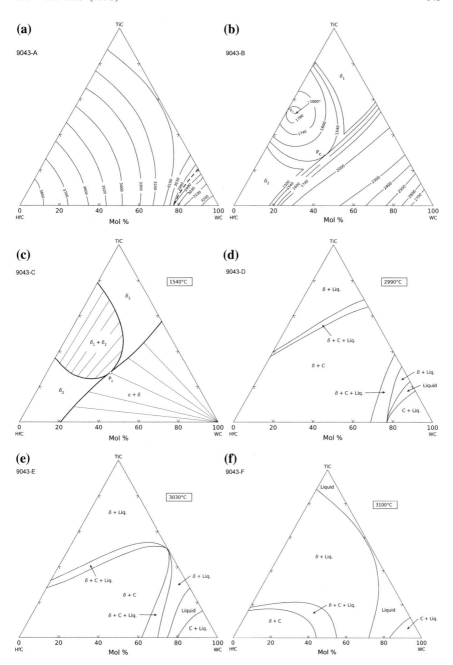

Fig. 4.37 **a** Liquidus projection of TiC–HfC–WC system. **b** Solidus phase boundaries at different temperatures. **c** Experimental isothermal sections at 1540 °C. **d** Experimental isothermal sections at 2990 °C. **e** Experimental isothermal sections at 3030 °C. **f** Experimental isothermal sections at 3100 °C. Reprinted with permission of The American Ceramic Society. **g** Isothermal section at 1500 °C comparing observed and calculated results. **h** Calculated isothermal section at 1600 °C and the calculated solubilities of WC in (Ti, Hf)C. Reprinted with permission of The American Ceramic Society

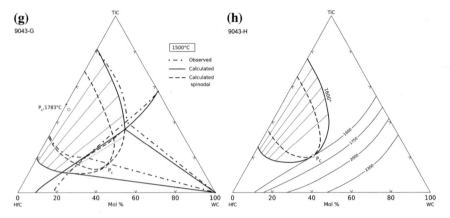

Fig. 4.37 (continued)

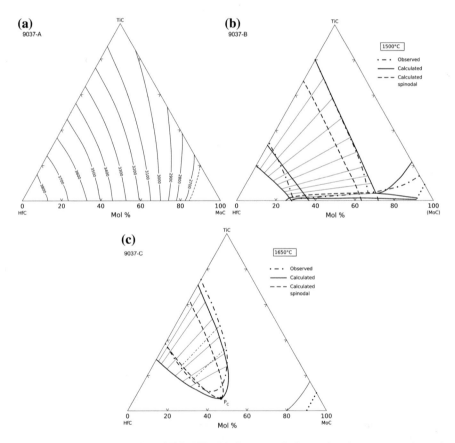

Fig.4.38 a Liquidus projection of TiC–HfC–(MoC) system. **b** Comparison between experimental and calculated isothermal data for the ternary miscibility gap at 1500 °C. **c** Comparison between experimental and calculated isothermal data for the ternary miscibility gap at 1650 °C (P_c = critical point on the binodal where the miscibility gap closes and separates from the solvus). Reprinted with permission of The American Ceramic Society

Sample preparation and the experimental procedures are the same as described in Sect. 4.38. The system was investigated using melting point studies [2]. Differential thermal analysis, X-ray diffraction, and metallographic techniques are performed from 1500 °C to the melting temperature. Lattice parameter values are given for various solid solution phases on the diagrams.

A pseudobinary miscibility gap forms within the HfC–MoC system below 1630 °C at $(HfC)_{0.45}(MoC)_{0.55}$. At lower temperatures within the pseudoternary system, a large miscibility gap connects the TiC–HfC and HfC–MoC boundary systems. Addition of MoC to the TiC–HfC solid solution has decreased the critical temperature, while addition of TiC to HfC–MoC compositions will raise the critical temperature. No maximum type ternary critical point was found.

Isothermal sections, Fig. 4.38b, c, were calculated assuming a regular solution model with spinodal–solvus interaction. The interaction parameters were obtained from the slope of experimental tie lines across the miscibility gap, the slope of the tangent at the solvus–binodal contact point, the maximum melting points of the solid solutions, and the critical evaluation of the metal–metal–carbon equilibria.

4.40   TiC–WC–NbC

Figure 4.39 shows the pseudoternary isothermal section of WC–TiC–NbC system at 1450 °C [51].

The system was studied as a Co-cemented carbide material containing 5–11 wt% Co (assumed non-reacting). An unstated number of compositions were studied by

Fig. 4.39 Pseudoternary isothermal section of WC–TiC–NbC system at 1450 °C. Reprinted with permission of The American Ceramic Society

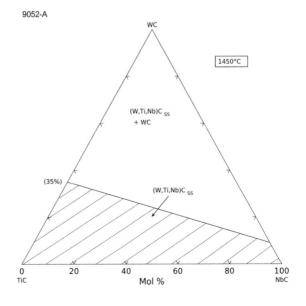

Fig. 4.40 Pseudoternary
isothermal section of WC–
TiC–TaC system at 1450 °C.
Reprinted with permission of
The American Ceramic
Society

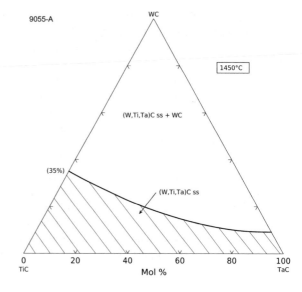

sintering in Ar at 1450 °C in the presence of Co, and cooling at 15 °C/min. The
products were examined by reflected light microscopy, microprobe analysis, and
X-ray diffraction. The solubility limit on the NbC–WC binary is in good agreement
with the extrapolation to 1450 °C of Rudy's data [4]. Lattice parameters are given
for several compositions of the gamma phase [(W, Ti, Nb)C].

4.41   TiC–WC–TaC

Figure 4.40 shows the pseudoternary isothermal section of WC–TiC–TaC system at
1450 °C [51].

Sample preparation and the experimental procedures are the same as described in
4.38. The solubility boundary on the TiC–WC binary is somewhat different from that
of Ref. [52], but is in good agreement with other Refs. [53–56]. Lattice parameters
are given for several compositions of the gamma phase γ [(W, Ti, Ta)C].

4.42   TiC–TaC/NbC–WC

The study was of carbide phase equilibrium in Co-based cemented carbide materials containing 5–11 wt% Co (assumed non-reacting). These diagrams were based
on the work shown in PED 9050, 9052, and 9055 on the pseudoternary systems,
and experimental work on nine compositions in the pseudoquaternary system. The
methods were the same as in the ternary systems. Lattice parameters are tabulated

Fig. 4.41 a Isothermal
section of TaC/NbC–WC–
TiC system with constant
TaC/NbC = 1.75 at 1450 °C.
b Isothermal section of TaC/
NbC–WC–TiC system with
constant TaC/NbC = 0.82 at
1450 °C. Reprinted with
permission of The American
Ceramic Society

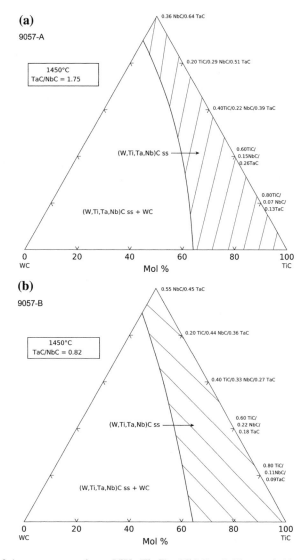

(a)

9057-A

1450°C
TaC/NbC = 1.75

0.36 NbC/0.64 TaC

0.20 TiC/0.29 NbC/0.51 TaC

0.40TiC/0.22 NbC/0.39 TaC

(W,Ti,Ta,Nb)C ss →

0.60TiC/
0.15NbC/
0.26TaC

0.80TiC/
0.07 NbC/
0.13TaC

(W,Ti,Ta,Nb)C ss + WC

0   20   40   60   80   100
WC   Mol %   TiC

(b)

9057-B

1450°C
TaC/NbC = 0.82

0.55 NbC/0.45 TaC

0.20 TiC/0.44 NbC/0.36 TaC

0.40 TiC/0.33 NbC/0.27 TaC

(W,Ti,Ta,Nb)C ss →

0.60 TiC/
0.22 NbC/
0.18 TaC

0.80 TiC/
0.11NbC/
0.09TaC

(W,Ti,Ta,Nb)C ss + WC

0   20   40   60   80   100
WC   Mol %   TiC

for several compositions of the γ gamma phase [(W, Ti, Ta, Nb)C ss]. Figure 4.41
shows an isothermal section of TaC/NbC–WC–TiC system with constant TaC/
NbC = 1.75 and 0.82 at 1450 °C [51].

4.43   HfC–NbC–VC

Figure 4.42 shows the isothermal section of NbC–HfC–VC system at 2050 °C
[49]. The point "*P*" is the calculated critical point.

The diagram is based on observations of ~15 ternary and 12 binary compositions. Starting materials were metal carbide powders prepared via carburization of either metal oxide or metal powders of unspecified purity at about 2000 °C. All compositions contained an additional 1 wt% Co to aid in solid-state diffusion rates. Mixtures were hot pressed and then annealed under hydrogen at temperatures for an additional 28 h to homogenize the solid solutions. Phase identification and lattice parameter determination were performed by X-ray diffraction. Lattice parameters of all solid solutions are presented graphically.

The critical point given in the figure was obtained through analysis of the solid solution data in terms of an ideal solution model. Reproduction of the experimentally determined tie lines with this model was poor, except as to the shape of the immiscibility region. Only the experimental data are reproduced here.

4.44   HfC–VC–WC

Figure 4.42 shows the pseudoternary phase equilibria of VC–HfC–WC system [57]. Figure 4.43a shows the liquidus projection. Figure 4.43b, c shows the experimental isothermal sections at 2800 and 2970 °C, respectively. Figure 4.43d shows the miscibility gap in the VC–HfC system. Figure 4.43e shows the comparison between experimental and calculated isothermal data at 2580 °C. The calculated isothermal sections at 1750, 2000, and 2300 °C are shown in Fig. 4.43f–h. The δ is the cubic monocarbide, (Hf, V, W)C$_{1-x}$ ss; δ_1, (V, Hf)C ss; δ_2, (Hf, V)C ss; P_c, critical point on the binodal where the miscibility gap closes and separates from the solvus.

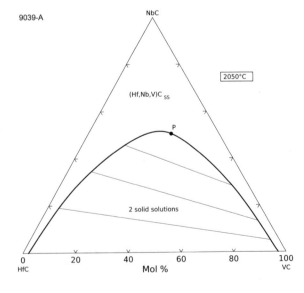

Fig. 4.42 Isothermal section of NbC–HfC–VC system at 2050 °C. Reprinted with permission of The American Ceramic Society

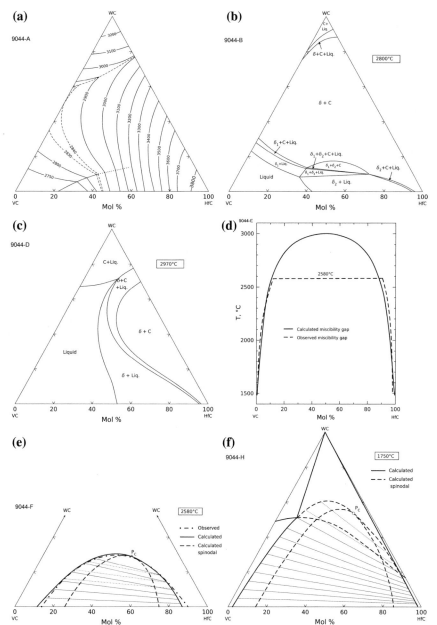

Fig. 4.43 a Liquidus projection of the pseudoternary phase equilibria in the VC–HfC–WC system. **b** Experimental isothermal sections at 2800 °C. **c** Experimental isothermal sections at 2970 °C. **d** Miscibility gap in the VC–HfC system. **e** Comparison between experimental and calculated isothermal data at 2580 °C. **f** Calculated isothermal sections at 1750 °C. **g** Calculated isothermal sections at 2000 °C. **h** Calculated isothermal sections at 2300 °C. Reprinted with permission of The American Ceramic Society

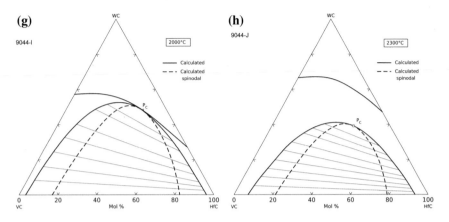

Fig. 4.43 (continued)

Sample preparation and the experimental procedures are the same as described in Sect. 4.39. A large pseudobinary miscibility gap forms within the VC–HfC system below ~3000 °C [PED9044 Figs. (E)–(G)]. Addition of WC decreases the critical temperature. The phase boundaries formed by interaction between the cubic monocarbide solid solution and the pseudoternary miscibility gap were studied at 2075 °C using $(VC)_{0.17}(HfC)_{0.37}(WC)_{0.46}$. Alloys in this region decompose upon cooling to give two cubic phases and hexagonal WC. Solid state and melting behavior were studied using studies of isopleths $VC_{0.88}$–WC, $VC_{0.88}$–$HfC_{0.98}$, $(V_{0.8}W_{0.2})C$–$(Hf_{0.8}W_{0.2})C$, and WC–$(V_{0.38}Hf_{0.62})C$.

Isothermal sections, Fig. (f–h), were calculated assuming a regular solution model with spinodal–solvus interaction. The interaction parameters were obtained from the slope of experimental tie lines across the miscibility gap, the slope of the tangent at the solvus–binodal contact point (2075 °C), the maximum melting points of the solid solutions, and the critical evaluation of the metal–metal–carbon equilibria.

Calculation of the $VC_{0.88}$–HfC section is given in Ref. [16] in which a model is proposed and combined with previously published thermochemical data to calculate the system at the high carbon boundary. The model is based on regular solution theory in which the interaction parameters are modified with respect to composition and temperature. The results of these calculations are compared with earlier experimental results [10, 11].

Editors' Note: A diagram described in the text as one comparing experimental and calculated data for an isothermal section at 1500 °C actually is such a diagram for the VC–HfC–MoC system at 2020 °C that appears in PED 9038.

4.45   HfC–VC–(MoC)

Figure 4.44 shows the phase diagram of VC–HfC–(MoC) system [58]. Figure 4.44a shows the liquidus projection. Figure 4.44b shows the experimental isothermal sections at 2700 °C. Figure 4.44c, d shows the calculated isothermal sections at 1630 and 2300 °C. Figure 4.44e, f shows the comparison between experimental and calculated isothermal data for the ternary miscibility gap at 2020 and 2700 °C. The δ, δ_1, and δ_2 are cubic monocarbide solid solutions where δ_1 and δ_2 are products of the spinodal decomposition of δss; the P_c is the critical point on the binodal where the miscibility gap closes and separates from the solvus.

The authors [58] emphasized that according to the melting behavior within the binary systems, this system does not represent a true pseudoternary system but for graphical presentation, it was considered as one. The liquidus surface in the "MoC"-rich corner, Fig. 4.44a, does not represent the crystallization of "MoC" but of C, see C–Mo system (PED 8943). The actual stoichiometries of the carbide end members are given as $VC_{0.88}$ and $HfC_{0.98}$.

Sample preparation and the experimental procedures are the same as described in Sects. 4.38 and 4.44. The system was investigated using melting point studies [2], differential thermal analysis, X-ray diffraction, and metallographic techniques from 1500 °C to the melting temperature. Lattice parameter values are given for various solid solution phases on diagrams in the paper.

A small pseudobinary miscibility gap forms within the HfC–MoC system below 1630 °C. Addition of VC to the HfC–MoC solid solution increases the critical temperature. Solid state and melting behavior were studied using the isopleths $VC_{0.88}$–MoC, $(V_{0.5}Hf_{0.5})C$–MoC, and $(V_{0.75}Hf_{0.25})C$–$(Hf_{0.75}Mo_{0.25})C$. Phase equilibria in the pseudoternary system are dominated by a large miscibility gap at 1500 °C.

Isothermal sections, Fig. 4.44c–f, were calculated assuming a regular solution model with spinodal–solvus interaction. The interaction parameters were obtained from the slope of the experimental tie lines across the miscibility gap, the slope of the tangent at the solvus–binodal contact point (2075 °C), the maximum melting points of the solid solutions, and the critical evaluation of the metal–metal–carbon equilibria. The slightly three-dimensional nature of Fig. 4.44f was not discussed but is probably related to the true quaternary nature of the system.

4.46   NbC–TaC–WC

Figure 4.45 shows the pseudoternary isothermal section of WC–NbC–TaC system at 1450 °C [51].

The system was studied as a co-cemented carbide material containing 5–11 wt% Co (assumed non-reacting). The boundary on the NbC–WC binary is in good agreement with the extrapolation to 1450 °C of Rudy's data [4]. Lattice parameters

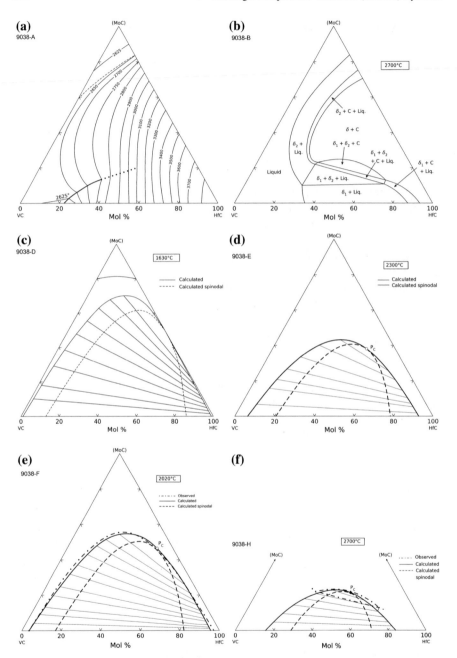

Fig. 4.44 a Liquidus projection of the VC–HfC–(MoC) system. **b** Experimental isothermal sections at 2700 °C. **c** Calculated isothermal sections at 1630 °C. **d** Calculated isothermal sections at 2300 °C. **e** Comparison between experimental and calculated isothermal data for the ternary miscibility gap at 2020 °C. **f** Comparison between experimental and calculated isothermal data for the ternary miscibility gap at 2700 °C. Reprinted with permission of The American Ceramic Society

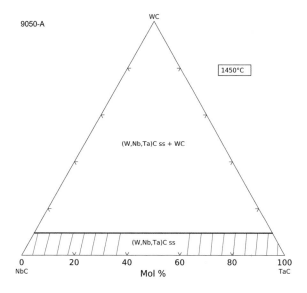

Fig. 4.45 Pseudoternary isothermal section of WC–NbC–TaC system at 1450 °C. Reprinted with permission of The American Ceramic Society

are given for several compositions of the γ gamma phase [(W, Ta, Nb)C]. There is some question as to the reliability of the data on the system because of the uncertainty of the composition of one of the samples.

4.47 V_2C–Ta_2C–W_2C

Figure 4.46 shows the isothermal sections of V_2C–Ta_2C–W_2C system at (a) 1650 °C and (b) 2000 °C [59].

The system was investigated using the element powders, W, Ta, V, and graphite, as starting materials to prepare the subcarbides. Heating then proceeded under He to 1300 °C to initiate reaction to form the subcarbides. Analysis showed that the Ta and Nb subcarbides were single phase and the V_2C showed some VC.

The subcarbides were then sintered at 1500–1850 °C for 2–3 h. The resulting subcarbide pellets were heated in a tantalum tube under He for 25 h at 2000 °C followed by 31 h at 1650 °C. The products of the treatment were examined by X-ray powder diffraction. Approximately, 35 compositions were examined at each temperature.

Figure 4.46a is taken from earlier reports [4, 60]. Lattice parameter data for the mixed metal carbides are presented graphically.

Fig. 4.46 a Isothermal
sections of the V$_2$C–Ta$_2$C–
W$_2$C system at 1650 °C.
b Isothermal sections of the
V$_2$C–Ta$_2$C–W$_2$C system at
2000 °C. Reprinted with
permission of The American
Ceramic Society

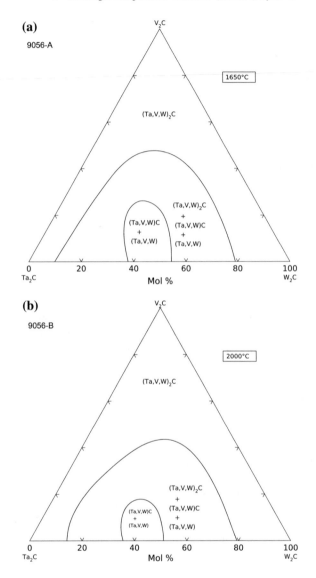

4.48   V$_2$C–Ta$_2$C–Mo$_2$C

Figure 4.47 shows the isothermal sections of V$_2$C–Ta$_2$C–Mo$_2$C system at
(a) 1650 °C and (b) 2000 °C [59]. Sample preparation and the experimental procedures are the same as described in Sect. 4.47. Figure 4.47a is taken from the
earlier reports [4, 60].

Fig. 4.47 a Isothermal
sections of the V$_2$C–Ta$_2$C–
Mo$_2$C system at 1650 °C.
b Isothermal sections of the
V$_2$C–Ta$_2$C–Mo$_2$C system at
2000 °C. Reprinted with
permission of The American
Ceramic Society

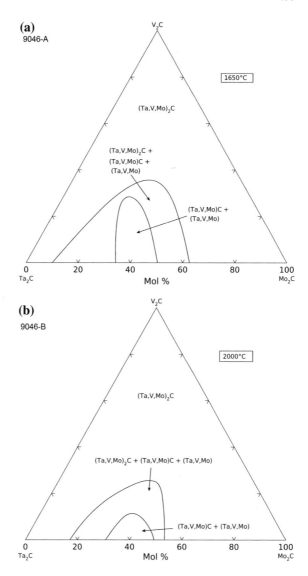

4.49   Nb$_2$C–Ta$_2$C–W$_2$C

Figure 4.48 shows the isothermal sections of Nb$_2$C–Ta$_2$C–W$_2$C system at
(a) 1650 °C and (b) 2000 °C [59]. Sample preparation and the experimental procedures are the same as described in Sect. 4.47.

Fig. 4.48 a Isothermal
sections of the Nb$_2$C–Ta$_2$C–
W$_2$C system at 1650 °C.
b Isothermal sections of the
Nb$_2$C–Ta$_2$C–W$_2$C system at
2000 °C. Reprinted with
permission of The American
Ceramic Society

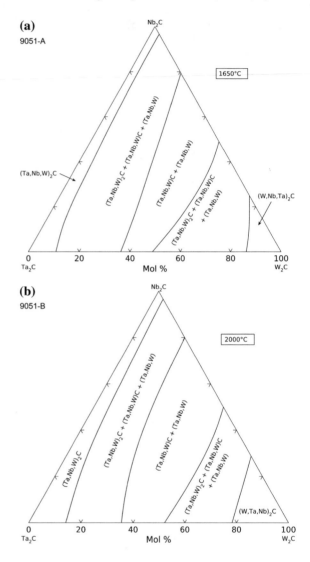

4.50   Nb$_2$C–Ta$_2$C–Mo$_2$C

Figure 4.49 shows the isothermal sections of Nb$_2$C–Ta$_2$C–Mo$_2$C system at
(a) 1650 °C and (b) 2000 °C [59]. Sample preparation is the same as described in
Sect. 4.47

Fig. 4.49 a Isothermal sections of the Nb₂C–Ta₂C–Mo₂C at 1650 °C system. **b** Isothermal sections of the Nb₂C–Ta₂C–Mo₂C at 2000 °C system. Reprinted with permission of The American Ceramic Society

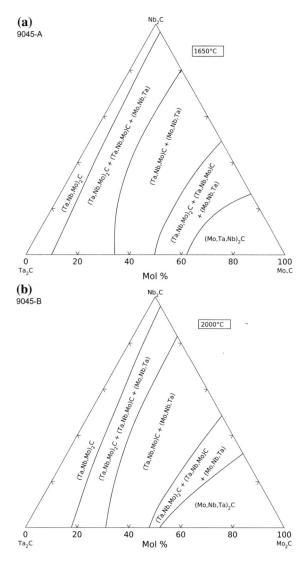

4.51   ZrC–ZrO₂–SiC–SiO₂–MgO

Figure 4.50 shows the phase diagram of SiC–SiO₂–ZrC–ZrO₂–MgO system at 1550 °C [61]. During the sintering of the SiC ceramics of the ternary SiC–MgO–ZrO₂ system, a solid reaction SiC + 2MgO + ZrO₂ ⟶ ZrC + Mg₂SiO₄ was found. Through the reaction, the ultrahigh-temperature ceramics ZrC was generated by common compounds. According to thermodynamics calculation, the values of the Gibbs free energies of this double substitution reaction are minus until 2000 °C. The ZrC was formed at 1250 °C through solid reaction, where the temperature was

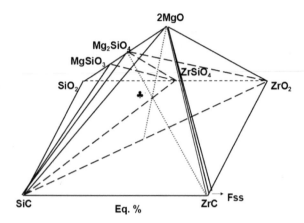

Fig. 4.50 Phase diagram of the SiC–SiO$_2$–ZrC–ZrO$_2$–MgO system at 1550 °C. Reprinted from Ref. [61], Copyright 2016, with permission from Elsevier

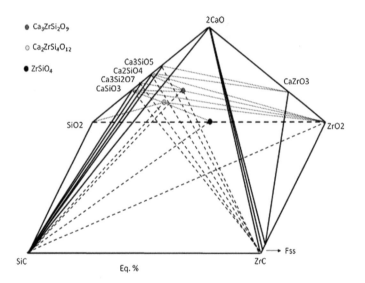

Fig. 4.51 Phase diagram of the SiC–SiO$_2$–ZrC–ZrO$_2$–CaO system at 1400 °C. Reprinted from Ref. [62], Copyright 2016, with permission from Elsevier

much lower than that of the carbon thermal reduction of ZrO$_2$. However, the tie line of the two compounds ZrC–Mg$_2$SiO$_4$ was not at the triangular facet of the ternary SiC–MgO–ZrO$_2$ system. It is across the centroid. Then, more experiments were conducted to confirm the phase relationship and the phase diagram of the quinary SiC–SiO$_2$–ZrC–ZrO$_2$–MgO system at 1550 °C (Fig. 4.50). The ZrC and SiC in the system were in equilibrium state with the three oxides and their silicates, and thus formed five four-phase coexistence tetrahedrons. The work provides a new route to the preparation of UHTC composites, such as ZrC/SiC, ZrC/silicate, by solid-state reaction at much lowered temperatures.

4.52   ZrC–ZrO₂–SiC–SiO₂–CaO

Figure 4.51 shows the phase diagram of SiC–SiO$_2$–ZrC–ZrO$_2$–CaO system at 1400 °C [62]. The same as that in the MgO-containing system (see Sect. 4.51), double substitution reaction SiC + 2CaO + ZrO$_2$ $\longrightarrow$ ZrC + Ca$_2$SiO$_4$ could happen at much lowered temperatures.

Phase relationships and the subsolidus phase diagram of the quinary system at 1400 °C were proposed and experimentally confirmed. ZrC and SiC coexist with four calcium silicates and two calcium zirconium silicates. CaZrO$_3$ coexists with ZrC only, and ZrSiO$_4$ coexists with SiC only. Therefore, 12 tetrahedrons of four-phase coexistence were formed.

The work provides a new route to the preparation of UHTC composites, such as ZrC/SiC and ZrC/silicate, by solid-state reaction at much lowered temperatures.

References

1. Rudy E, Windisch S (1966) Part II. Ternary systems, vol XIII, Phase diagrams of the systems Ti-B-C, Zr-B-C, and Hf-B-C, ternary phase equilibria in transition metal-boron-carbon-silicon systems. Air Force Materials Laboratory; Wright-Patterson Air Force Base, Ohio
2. Rudy E, Progulski J (1967) A Pirani furnace for the precision determination of the melting temperatures of refractory metallic substances. Planseeber Pulvermetallurgie 15(1):13–45
3. Kirfel A, Gupta A, Will G (1979) The nature of the chemical bonding in boron carbide, B$_{13}$C$_2$. I. Structure refinement. Acta Crystallogr B 35 (5):1052–1059
4. Rudy E (1969) Part V. Compendium of phase diagram data, 689 pp. Ternary phase equilibria in transition metal-boron-carbon-silicon systems. Air Force Materials Laboratory; Wright-Patterson Air Force Base, Ohio
5. Ordan'yan SS, Unrod VI, Avgustinik AI (1975) Reactions in the system TiC$_x$-TiB$_2$. Soviet powder metallurgy and metal ceramics 14(9):729–731
6. Beratan HR (1980) The directional solidification and properties of the TiC-TiB$_2$ eutectic. MS thesis, Pennsylvania State University, Philadelphia
7. Gusev AI (1984) Calculation of the phase diagrams of pseudobinary systems based on high-heat titanium, zirconium, hafnium, and vanadium carbides. Izvestiya Akademii Nauk SSSR, Neorganicheskie Materialy 20(7):1132–1137. Inorg Mater (English Translation), 20(7):976–981
8. Hardy HK (1953) A "sub-regular" solution model and its application to some binary alloy systems. Acta Metall 1(2):202–209
9. Gusev AI (1985) Formation of solid solutions by transition-metal carbides and nitrides and calculation of their phase diagrams. In: Akad. Wiss. DDR, Zentralinst. Grundlagen, Herstell. Eigenschaften Pulvermetall. Werkst. Dresden: Intern. Pulvermet. Tagung. DDR, 8th 3:169–179
10. Kieffer R, Nowotny H, Neckel A et al (1968) Zur entmischung von kubischen mehrstoffcarbiden. Monatshefte für Chemie-Chemical Monthly 99(3):1020–1027
11. Kieffer R (1969) Preparation and properties of interstitial compounds. J Inst Met 97(6):164–172
12. Dmitrieva GP, Razumova NA, Shurin AK (1984) Phase equilibrium diagram of the Ni-TiC-HfC system. Sov Powder Metall Met Ceram 23(2):159–162

13. Rogl P, Naik SK, Rudy E (1977) A constitutional diagram of the system TiC–HfC–WC. Monatsh Chem 108(5):1189–1211
14. Brukl CE, Harmon DP (1966) The Ti-Zr-C, Ti-Hf-C and Zr-Hf-C systems. Technical. ternary phase equilibria in transition metal-boron-carbon-silicon systems, part II. In: Ternary systems, vol IV. Air Force Materials Laboratory, Wright-Patterson Air Force Base, Ohio, pp 1–78
15. Gusev AI (1985) Phase diagrams of the pseudo-binary TiC-NbC, TiC-TaC, ZrC-NbC, ZrC-TaC, and HfC-TaC carbide systems. Russ J Phys Chem (English Translation) 59(3):336–340
16. Rempel AA, Gusev AI, Shveikin GP (1984) Thermodynamic calculation of the phase diagrams of the $VC_{0.88}$-NbC, $VC_{0.88}$-TaC, $VC_{0.88}$-HfC, NbC-TaC, and NbC-HfC systems. Russ J Phys Chem (English Translation) 58(9):1322–1325
17. Ordan'Yan SS, Nikolaeva EE (1987) Interaction in the GdB_6-TiB_2 system. Sov Powder Metall Met Ceram (English Translation) 26(1):51–53
18. Chaban NF, Kuz'ma YB, Gerasim ID (1978) {Ti, Zr, Hf}-Gd-B systems. Sov Powder Metall Met Ceram (English Translation) 17(8):592–593
19. Samsonov GV, Serebryakova TI, Neronov VA (1975) Borides. Atomizdat, Moscow, USSR, p 375 (in Russian)
20. Blanks JH (1979) Dissertation abstract international B. Dissertation, 40(1):402B–402B
21. Hawkins DT, Hultgren R (1973) Constitution of binary alloys. In: Taylor L (ed) Metals handbook, vol 8. American Society of Metals. Metals Park, Ohio p, pp 251–376
22. Chupov VD, Ordan'yan SS, Kozlovskii LV (1981) Reaction in the TiN_x-TiB_2 system. Inorg Mater 17(9):1195–1198
23. Rogl P, Schuster JC (1992) Phase diagrams of ternary boron nitride and silicon nitride systems. In: Monograph series on alloy phase diagrams. ASM International, Materials Park, Ohio Boron Nitride & Silicon Nitride Systems, pp 103–106
24. Ordan'yan SS, Unrod VI (1975) Reactions in the system ZrC-ZrB_2. Sov Powder Metall Met Ceram (English Translation) 14(5):393–395
25. Gusev AI (1985) Prediction and calculation of phase diagrams of pseudobinary systems based on high-melting transition metal compounds. In: Ageev Izd NV (ed) Raschety Eksp. Metody Postroeniya Diagramm Sostoyaniya. Moscow, USSR, p 42–47
26. Ordan'yan SS, Dmitriev AI, Moroshkina ES (1989) Reaction of SiC with ZrB_2. Inorg Mater (English Translation) 25(10):1752–1755
27. Hansen M, Anderko K (1958) Constitution of binary alloys, 2nd edn. In: Mehl RF, Bever MB. McGraw-Hill Companies, Inc., New York, pp 378–380
28. Ordan'yan SS (1993) Obschie aspekty fazovykh otnosheniy v sistemakh SiC-MeIV-VIB₂, Zh. Prikl. Khim. (in Russian), 66(11):2439–2444. Common aspects of phase relations in SiC-MeIV-VIB2 system. J Appl Chem USSR (English Translation), 66(11):1855–1858
29. Ordan'yan SS, Chupov VD (1984) Interaction in ZrN-ZrB_2 and HfN-HfB_2 systems. Inorg Mater (English Translation) 20(12):1719–1722
30. Eron'yan MA, Avarbe RG. Nikol'skaya TA (1976) How the pressure of nitrogen affects the melting point of ZrN_x (title of translated paper). Neorganicheskie Materialy 12(2):247–249
31. Chang YA (1966) Part II. Ternary systems, vol IX. Zr-W-B system and the pseudobinary system TaB2-HfB2, Ternary phase equilibria in transition metal-boron-carbon-silicon systems. Report No. AFML-TR-65-2, Contract No. USAF 33(615)-1249.Air Force Materials Laboratory; Wright-Patterson Air Force Base, Ohio, pp 1–26
32. Romans PA, Krug MP (1966) Composition and crystallographic data for the highest boride of tungsten. Acta Crystallogr A 20(2):313–315
33. Post B, Glaser FW (1952) Crystal structure of ZrB_{12}. Trans Am Inst Min Metall Eng 194 (6):631–632
34. Ordan'yan SS, Nikolaeva EE, Martynova TY (1987) Interaction in the system GdB_6-HfB_2. Russ J Inorg Mater (English Translation) 32(11):1665–1666
35. Harmon DP (1966) Part II. Ternary systems, vol XI. Hf-Mo-B and Hf-W-B Systems, Ternary phase equilibria in transition metal-boron-carbon-silicon systems. Report

No. AFML-TR-65-2, Contract No. USAF 33(615)-1249. Air Force Materials Laboratory; Wright-Patterson Air Force Base, Ohio, pp 1–41

36. Rogl P, Nowotny H, Benesovsky F (1971) Komplexboride in den systemen Hf-Mo-B und Hf-W-B. Monatshefte für Chemie-Chemical Monthly 102(4):971–984

37. Khan YS, Kalmykov KB, Dunaev SF et al (2004) Phase equilibria in the Ti-Al-N system at 1273 K. Dokl Phys Chem (English Translation) 396(2):134–137

38. Abramycheva NL, Kalmikov KB, Dunaev SE (2002) Application of a method of high-temperature nitriding for research of interaction nitride and intermetallides phases in systems with participation of nitrogen. Perspektivnye Materialy 5:83–89

39. Schuster JC, Bauer J, Nowotny H (1985) Applications to materials science of phase diagrams and crystal structures in the ternary systems of transition metal-Al-N. Rev Chim Miner 22 (4):546–554

40. Khan YS, Kalmykov KB, Dunaev SF et al (2004) Phase equilibria in the Zr-Al-N system at 1273 K. Metally (Moscow), pp 54–63

41. Ma Y, Romming C, Lebech B et al (1992) Structure refinement of Al_3Zr using single-crystal X-ray diffraction, powder neutron diffraction and CBED. Acta Crystallogr B: Struct Sci Cryst Eng Mater 48(1):11–16

42. Lu YJ, Huai XC, Wu LE et al (2015) Phase composition of ZrN-Si_3N_4-Y_2O_3 composite material. J Chin Ceram Soc 43(12):1742–1746

43. Wu LE, Sun WZ, Chen YH et al (2011) Phase relations in Si-C-N-O-R (R = La, Gd, Y) systems. J Am Ceram Soc 94(12):4453–4458

44. Shurin AK, Dmitrieva GP, Razumova NA et al (1987) The phase diagram of the Ni-VC-NbC system. Sov Powder Metall Met Ceram 26(8):658–660

45. Ordan'yan SS, Dmitriev AI, Bizhev KT et al (1987) Interaction in B_4C-Me_vB_2 systems. Powder Metall Met Ceram 26(10):834–836

46. Ordan'yan SS, Topchii LA, Khoroshilova IK et al (1982) Reactions in the $VC_{0.88}$-VB_2 system. Sov Powder Metall Met Ceram 21(2(230)):122–124

47. Rudy E (1970) "Part V. The phase diagram W-B-C", Experimental phase equilibria of selected binary, ternary, and higher order systems. Air Force Materials Laboratory; Wright-Patterson Air Force Base, Ohio

48. Rudy E, Benesovsky F, Toth L (1963) Studies of the ternary systems of the group Va and VIa metals with boron and carbon. Zeitschrift für Metallkunde 54(6):345–353

49. Rudy E, Nowotny H, Benesovsky F et al (1960) Über Hafniumkarbid enthaltende Karbidsysteme. Monatsh für Chem-Chem Monthly 91(1):176–187

50. Rogl P, Naik SK, Rudy E (1977) A constitutional diagram of the system TiC-HfC-"MoC". Monatsh für Chem-Chem Monthly 108(6):1325–1337

51. Chatfield C (1986) The gamma/WC solubility boundary in the quaternary TiC-NbC-TaC-WC system at 1723 K. J Mater Sci 21(2):577–582

52. Chatfield C (1983) A redetermination of the gamma-alpha solubility line in the TiC-TaC-WC system at 1723 K. Int J Powder Metall 15(1):18–19

53. Uhrenius B (1984) Calculation of the Ti-C, W-C and Ti-W-C phase diagrams. CALPHAD: Comput Coupling Phase Diagrams Thermochem 8(2):101–119

54. Rudy E (1973) Constitution of ternary titanium-tungsten-carbon alloys. J Less Common Met 33(2):245–273

55. May W (1972) Zum Reaktionsverhalten übersättigter (Wolfram, Titan, Tantal)-Karbide in flüssigem Wolfram. Techn. Mitt. Krupp 30(1):15–48

56. Chaporova IN, Cheburaeva RF (1973) Precise determination of the boundaries of the phase regions in the pseudo ternary TiC-WC-TaC system. Naucnye trudy VNIITS 12:78–83

57. Rogl P, Naik SK, Rudy E (1977) A constitutional diagram of the system $VC_{0.88}$-$HfC_{0.98}$-WC. Monatsh für Chem-Chem Monthly 108(5):1213–1234

58. Rogl P, Naik S K, Rudy E (1977) A constitutional diagram of the system $VC_{0.88}$-$HfC_{0.98}$-(MoC). Monatsh für Chem-Chem Monthly 108(6):1339–1352

59. Brukl CE (1972) Der Einfluß von Vanadin und Niob auf feste Subcarbidlösungen in den Systemen Tantal-Wolfram-Kohlenstoff und Tantal-Molybdän-Kohlenstoff. Monatsh für Chem-Chem Monthly 103(3):820–830
60. Brukl CE (ed) (1969) Part IV. The effect of molybdenum and tungsten on the subcarbide solutions in the vanadium-tantalum-carbon and niobium-tantalum-carbon systems, phase equilibria investigation of binary, ternary, and higher order systems. Air Force Materials Laboratory; Wright-Patterson Air Force Base, Ohio, pp 1–44
61. Sun WZ, Cheng JG, Huang ZK et al (2016) ZrC formation and the phase relations in the Si-Zr-Mg-O-C system. J Mater Sci 51(17):8139–8147
62. Sun WZ, Cheng JG, Huang ZK et al (2016) ZrC formation and the phase relations in the SiC-SiO$_2$-ZrC-ZrO$_2$-CaO system. Ceram Int 42(8):10165–10170

Erratum to: Phase Equilibria Diagrams of High Temperature Non-oxide Ceramics

Erratum to:
Z. Huang and L. Wu, *Phase Equilibria Diagrams*
of High Temperature Non-oxide Ceramics,
https://doi.org/10.1007/978-981-13-0463-7

The original version of the book was inadvertently published without incorporating the belated corrections from author, which have to be now incorporated. The erratum book has been updated with the changes.

The updated online version of this chapters can be found at
https://doi.org/10.1007/978-981-13-0463-7_3
https://doi.org/10.1007/978-981-13-0463-7_4
https://doi.org/10.1007/978-981-13-0463-7

© Springer Nature Singapore Pte Ltd. 2018
Z. Huang and L. Wu, *Phase Equilibria Diagrams of High Temperature*
Non-oxide Ceramics, https://doi.org/10.1007/978-981-13-0463-7_5

Appendices

See Tables A.1, A.2, A.3, A.4, A.5, and A.6.

Table A.1 Melting point and structure of some compounds in Y–Si–Al–O–N system

Composition	Melting point (°C)	Structure	Notes
Si_3N_4, α-, β-SiAlON	1900	Hexagonal	Stable in inert atmosphere
Si_2N_2O, O′-SiAlON	1900	Orthorhombic	Good oxidation resistance
AlN	2450	Hexagonal, cubic	Stable in inert atmosphere
YN	∼3000	Cubic (NaCl-type)	Estimated melting point
AlN Polytypoids	>2200	nH:hexag.nR: rhom.	Stable in inert atmosphere
Nd_2AlO_3N	1750	Tetragonal	Congruent melt
$NdAl_{12}O_{18}N$	1850	β-Al_2O_3-type	The highest stable temp.
$5AlN·9Al_2O_3$	2165	Cubic	AlON(γ)
$2YN·Si_3N_4$	>1800	Orthorhombic	Synthesized at 1800 °C
X-phase	1727	Tric., Monoc.,Ortho.	Mullite-type
$Y_2O_3·Si_3N_4$	1825	Tetrag. (M-phase)	Y-melilite
$Y_2O_3·Si_2N_2O$	∼1900	Hexag. (K-phase)	Y-wollastonite
$2Y_2O_3·Si_2N_2O$	∼2000	Monocl. (J-phase)	Y-cuspidine
$5Y_2O_3·4SiO_2·Si_2N_2O$	1600 (La)	Hexag. (H-phase)	Y(La)-apatite
Y_2O_3	2400	Cubic	Congruent melt
Al_2O_3	2050	Hexag. Corundum	Congruent melt
SiO_2	1725	Cristobalite	Congruent melt
$2Y_2O_3·Al_2O_3$	2020	Monoclinic (YAM)	Woehlerite type
$Y_2O_3·Al_2O_3$	1400–1875	Perovskite (YAP)	Stable temperature range
$3Y_2O_3·5Al_2O_3$	1930	Cubic (YAG)	Y-garnet
$Y_2O_3·SiO_2$	1959 (calculated)	Monoclinic	Congruent melt calculated
$2Y_2O_3·3SiO_2$	1630–1920	Tetragonal	Stable temperature range
$Y_2O_3·2SiO_2$	1787 (calculated)	Monoclinic	Incongruent melt

© Springer Nature Singapore Pte Ltd. 2018
Z. Huang and L. Wu, *Phase Equilibria Diagrams of High Temperature Non-oxide Ceramics*, https://doi.org/10.1007/978-981-13-0463-7

Table A.2 Eutectic temperature and composition of some N-ceramic systems

System	Teu., °C	Eutectic composition
Si_3N_4–SiO_2	1593	Near SiO_2 side
Si_3N_4–La_2O_3	~1600	Close to 20 Si_3N_4/80 La_2O_3 (m%)
Si_3N_4 –Y_2O_3	1720	$15Si_3N_4 + 85Y_2O_3$ (m%)
Si_3N_4 –$Y_3Al_5O_{12}$(YAG)	1650	$25Si_3N_4 + 75$YAG (w%)
Si_3N_4 - $Y_2Si_2O_7$	1570	$10Si_3N_4 + 90Y_2Si_2O_7$ (w%)
YAG–$Y_2Si_2O_7$	1520	35YAG $+ 65Y_2Si_2O_7$ (w%)
Si_3N_4–YAG–$Y_2Si_2O_7$	1430	$10Si_3N_4 + 27$YAG $+ 63Y_2Si_2O_7$ (w%)
Si_3N_4–Mg_2SiO_4	1560	$7Si_3N_4 + 93Mg_2SiO_4$ (m%)
Si_3N_4–SiO_2–MgO–CaO	1325	$11Si_3N_4 + 34SiO_2 + 22$MgO $+ 33$CaO (m%)
Si_3N_4–SiO_2–Al_2O_3–AlN	1480	Close to rich-SiO_2 area
Si_3N_4–SiO_2–La_2O_3	<1650	Glass area near $75La_2O_3$/$25Si_3N_4$ (m%)
Si_3N_4–AlN–Y_2O_3	1650	$15Si_3N_4 + 25$AlN $+ 60Y_2O_3$ (m%)
Si_2N_2O–Al_2O_3–Y_2O_3	1450	Close to O′-SiAlON-Al_2O_3 line
SiO_2–Al_2O_3–Y_2O_3	1360	Close to rich-SiO_2 area
Si_3N_4–Si_2ON_2–Mg_2SiO_4	1510 (calculated)	$0.153Si_3N_4 + 0.194Si_2ON_2 + 0.653Mg_2SiO_4$
Si_3N_4–Si_2ON_2–Mg_2SiO_4	1515 (experim.)	$4Si_3N_4 + 14Si_2ON_2 + 82Mg_2SiO_4$ (m%)
Si_2ON_2–Mg_2SiO_4	1525	$15Si_2ON_2 + 85Mg_2SiO_4$ (m%)
AlN–La_2O_3	~1650	Near rich-La_2O_3 side
AlN–Y_2O_3	1730	20AlN $+ 80Y_2O_3$ (m%)
AlN–Nd_2O_3	1660	30AlN $+ 70Nd_2O_3$ (m%)
AlN–Eu_2O_3	1375	55AlN $+ 45Eu_2O_3$ (m%)

Table A.3 Compounds of nitrogen-containing metal aluminosilicates

Formulae	Components	Structures
LiSiON	$Li_2O \cdot Si_2N_2O$	Hexagonal Wurtzite
$LiAlSiON_2$	$Li_2O \cdot 2AlN \cdot Si_2N_2O$	Orthorhombic
$LiAlSiO_{4-x}N_x$	$Li_2O \cdot Al_2O_3 \cdot 2SiO_2$ (xN Subst.xO)	β-eucryptite
$LiSi_2N_3$	$Li_3N \cdot 2Si_3N_4$	Orthorhombic
$Li_{14}Si_9O_7N_{12}$	$7Li_2O \cdot 3Si_3N_4$	Unknown
$MgAl_2Si_4O_6N_4$	$MgO \cdot Al_2O_3 \cdot 2Si_2N_2O$	Orthorhombic (N)
$MgAlSiN_3$	$Mg_3N_2 \cdot 3AlN \cdot Si_3N_4$	Orthorhombic
$MnAlSiN_3$	$Mn_3N_2 \cdot 3AlN \cdot Si_3N_4$	Orthorhombic
$CaAlSiN_3$	$Ca_3N_2 \cdot 3AlN \cdot Si_3N_4$	Orthorhombic (E)
$Ca_3Si_2O_4N_2$	$3CaO \cdot Si_2N_2O$	Cubic (Z)
$CaSi_2O_2N_2$	$CaO \cdot Si_2N_2O$	Unknown

(continued)

Table A.3 (continued)

Formulae	Components	Structures
$Ca_2Si_3O_2N_4$	$2CaO \cdot Si_3N_4$	Unknown
$Ca_2AlSi_3O_2N_5$	$2CaO \cdot AlN \cdot Si_3N_4$	Metastable
$CaO \cdot 1.33Al_2O_3 \cdot 0.67Si_2N_2O$	$Ca_3Al_8Si_4O_{17}N$	Unknown (S)
$Ca_3Al_2 (Si_3O_6N_4)$	$3CaO \cdot Al_2O_3 \cdot Si_3N_4$	Cubic Garnet
$Sr_2Al_xSi_{12-x}O_{2+x}N_{16-x}$	$SrO \cdot Si_5OAlN_7$ $(x = 2)$	Unknown
$Al_{23}O_{27}N_5$	$9Al_2O_3 \cdot 5AlN$	Cubic AlON (γ)
$Al_6Si_6O_9N_8$	$3Al_2O_3 \cdot 2Si_3N_4$	Mullite-type (X)
$Al_{18}Si_{12}O_{39}N_8$	$9Al_2O_3 \cdot 6SiO_2 \cdot 2Si_3N_4$	Mullite-type (X)

Table A.4 Formulae and structures of some nitrogen-containing rare earth aluminosilicates

Formulae	Components	Structures
$R_2Si_6O_3N_8$	$R_2O_3 \cdot 2Si_3N_4$ (R = La, Ce)	Monoclinic
$R_2Si_3O_3N_4$	$R_2O_3 \cdot Si_3N_4$ (R = Ce, Nd, Sm, Gd, Dy, Y, Er, Yb)	Tetragonal melilite (M)
$RSiO_2N$	$R_2O_3 \cdot Si_2N_2O$ (R = La, Ce, Nd, Sm, Y)	Hexagonal Wollastonite (K)
$R_5(SiO_4)_3N$	$5R_2O_3 \cdot 4SiO_2 \cdot Si_2N_2O$ (R = La,Ce,Nd,Sm,Gd, Dy,Y,Er,Yb)	Hexagonal apatite (H)
$R_4Si_2O_7N_2$	$2R_2O_3 \cdot Si_2N_2O$ (R = La, Ce, Pr, Nd, Sm, Gd, Tb, Dy, Y, Ho, Er, Tm, Yb and Lu)	Monoclinic Cuspidine (J)
$RAl_{12}O_{18}N$	$RN \cdot 6Al_2O_3$ (R = La, Ce, Nd, Sm)	Hexagonal Magnetoplumbite
$R_3Si_{8-x}Al_xN_{11-x}O_{4+x}$	$0 < x < 1.75$, (R = La, Nd, Ce)	Monoclinic
$R_4Si_9Al_5O_{30}N$	$2R_2O_3 \cdot 2Al_2O_3 \cdot 9SiO_2 \cdot AlN$ (R = La, Ce, Nd)	Monoclinic
R_2SiAlO_5N	$R_2O_3 \cdot SiO_2 \cdot AlN$ (R = Dy, Y, Er, Yb)	Monoclinic
$R_3Si_{3-x} Al_{3+x}O_{12+x} N_{2-x}$ $(0 < x < 1)$	$3R_2O_3 \cdot 3Al_2O_3 \cdot 2SiO_2 \cdot 2Si_2N_2O$ (R = La, Ce, Nd, Sm, Gd, Dy, Y)	Gallium Germanate-type Ga $(R_3Ga_5GeO_{14})$
R_2AlO_3N	$R_2O_3 \cdot AlN$ (R = Ce, Pr, Nd, Sm, Eu)	Tetragonal
$La_2Si_2AlO_4N_3$	$La_2O_3 \cdot Si_2N_2O \cdot AlN$	Tetragonal La-melilite
$LaEuSiO_3N$	$La_2O_3 \cdot 2EuO \cdot Si_2N_2O$	β-K_2SO_4-type
$La_6Y_4Si_{45}O_{15}N_{60}$	$3La_2O_3 \cdot 2Y_2O_3 \cdot 15Si_3N_4$	unknown
$RAl(Si_{6-z}Al_z)$ $(N_{10-z}O_z)$ $(z<1)$	$RN \cdot AlN \cdot 2Si_3N_4$ $(z = 0)$, (R = La, Nd)	Unknown (Jem)
$SrSi_7N_{10}$	$Sr_3N_2 \cdot 7Si_3N_4$	Monoclinic
$Y_2Si_3N_6$	$2YN \cdot Si_3N_4$	Orthorhombic
Y_2AlSiO_5N	$2Y_2O_3 \cdot Al_2O_3 \cdot Si_2N_2O$	Metastable
$Y_3AlSi_2O_7N_2$	$3Y_2O_3 \cdot Al_2O_3 \cdot 2Si_2N_2O$	Pseudohexagonal
$MRSi_4N_7$	M = Ca, Sr, Ba, Eu; R = Y, La, Yb	Hexagonal
$R_2Si_3AlN_7$	R = Y, La or Y + La	Hexagonal
$(Ca,Y)_2Si_4(N,C)_7$	$Ca_{0.8}Y_{1.2}Si_4N_{6.8}C_{0.2}$	Hexagonal
$R_2Si_4N_6C$	R = Y or La	Orthorhombic

Table A.5 Crystal chemistry data of some compounds with tetrahedral [MX₄] structure

	SiO₂	Si₂N₂O	Si₃N₄	SiC	AlN	BeO	Al₂O₃
Structure[a]	H	O	H	H(α) C(β)	H, C	H	H
Structural elements	[SiO₄]	[SiN₃O]	[SiN₄]	[SiC₄]	[AlN₄]	[BeO₄]	{AlO₆}[b]
Coordination number	4:2	4:3:2	4:3	4:4	4:4	4:4	6:2
Bond length (Å)	1.62	1.72 (Si–N)	1.74	1.89	1.87	1.65	1.90(6) (1.75{4})
Bond ionicity	0.5	0.3 (Si–N) 0.5 (Si–O)	0.3	0.2	0.4	0.5	0.6

[a]H hexagonal, O rhombohedral, C cubic
[b]Al–O Coordination number in Al₂O₃ is [AlO₆] octahedron

Table A.6 Crystal structure, lattice parameter, density, and melting point of Ultrahigh-Temperature Materials (UHTMs)

Formula	Crystal structure	Lattice parameters			Density (g/ml)	Melting point (°C)
		a (Å)	b (Å)	c (Å)		
HfB₂	Hexag.	3.142	–	3.476	11.19	3380
HfC	FCC	4.638	4.638	4.638	12.76	3900
HfN	FCC	4.525	4.525	4.525	13.9	3385
ZrB₂	Hexag.	3.169	–	3.530	6.10	3245
ZrC	FCC	4.693	4.693	4.693	6.56	3400
ZrN	FCC	4.578	4.578	4.578	7.29	3200[a]
TiB₂	Hexag.	3.030	–	3.230	4.52	3225
TiC	FCC	4.327	4.327	4.327	4.94	3100
TiN	FCC	4.242	4.242	4.242	5.39	3050[a]
ThO₂	FCC	5.597	5.597	5.597	10.00	3390
TaB₂	Hexag.	3.098	–	3.227	12.54	3040
TaC	FCC	4.455	4.455	4.455	14.50	3800
NbB₂	Hexag.	3.160		3.400	6.97	3000
NbC	FCC	4.470	4.470	4.470-	7.820	3610
W	BCC	–	–	–	19.25	3410
C	FCC	–	–	–	3.515	3550
Re	Hexag	–	–	–	21.02	3200
Os	Hexag	–	–	–	22.59	3030
Ta	BCC	–	–	–	16.69	3020

[a]Melting point depends on the nitrogen pressure, see Chap. 4, Ref. [30]

Printed in the United States
By Bookmasters